INSECT
ANATOMY

THE CURIOUS WORLD OF BEES, BEETLES, BUTTERFLIES, AND BUGS

JULIA RoTHMAN
BEST-SELLING AUTHOR OF THE ANATOMY SERIES
& MICHAEL HEARST

The mission of Storey Publishing is to serve our customers by
publishing practical information that encourages
personal independence in harmony with the environment.

The information in this book is true and
complete to the best of our knowledge. All
recommendations are made without guarantee
on the part of the author or Storey Publishing.
The author and publisher disclaim any liability in
connection with the use of this information.

The publisher is not responsible for websites
(or their content) that are not owned by the
publisher.

Storey books may be purchased in bulk for
business, educational, or promotional use.
Special editions or book excerpts can also
be created to specification. For details,
please contact your local bookseller or the
Hachette Book Group Special Markets
Department at special.markets@hbgusa.com.

Storey Publishing
210 MASS MoCA Way
North Adams, MA 01247
storey.com

Storey Publishing is an imprint of Workman
Publishing, a division of Hachette Book Group,
Inc., 1290 Avenue of the Americas, New York,
NY 10104. The Storey Publishing name and logo
are registered trademarks of Hachette Book
Group, Inc.

ISBNs: 978-1-63586-878-4 (paperback);
978-1-63586-879-1 (fixed format EPUB);
978-1-63586-999-6 (fixed format Kindle);
979-8-89708-004-5 (fixed format PDF)

Printed in China by R. R. Donnelley on paper
from responsible sources
10 9 8 7 6 5 4 3 2

APS

Library of Congress Cataloging-in-Publication
Data on file

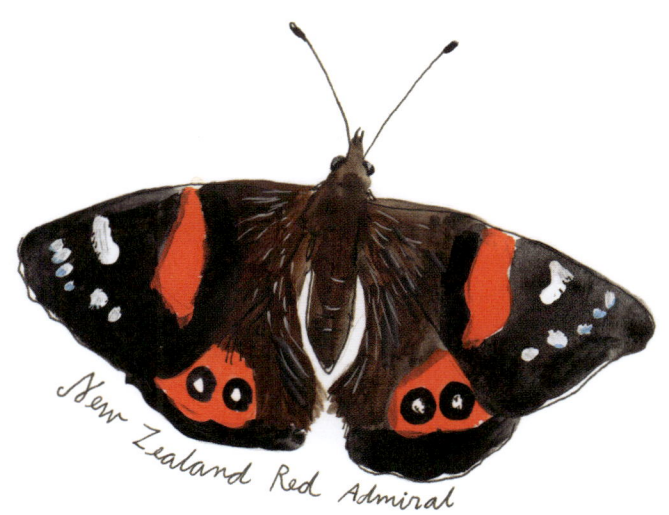

New Zealand Red Admiral

FOR MY NEW, NEW ZEALAND FRIENDS:

JEREMY, CHRIS, PAULINE, JOHN, MIRANDA, AMANDA, WILBUR, AMY, GUY, SANTASHREE, AKASH, LUKA, RIJEKA, ROWAN

CONTENTS

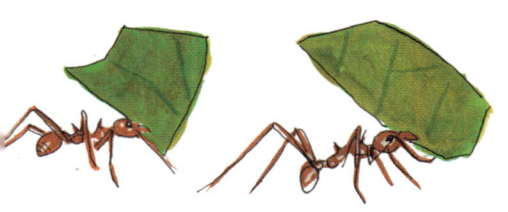

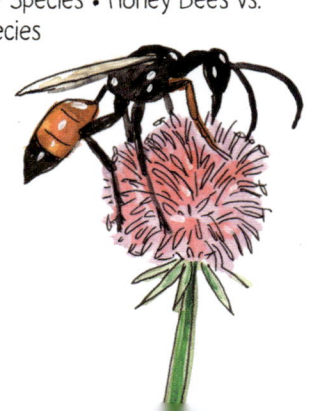

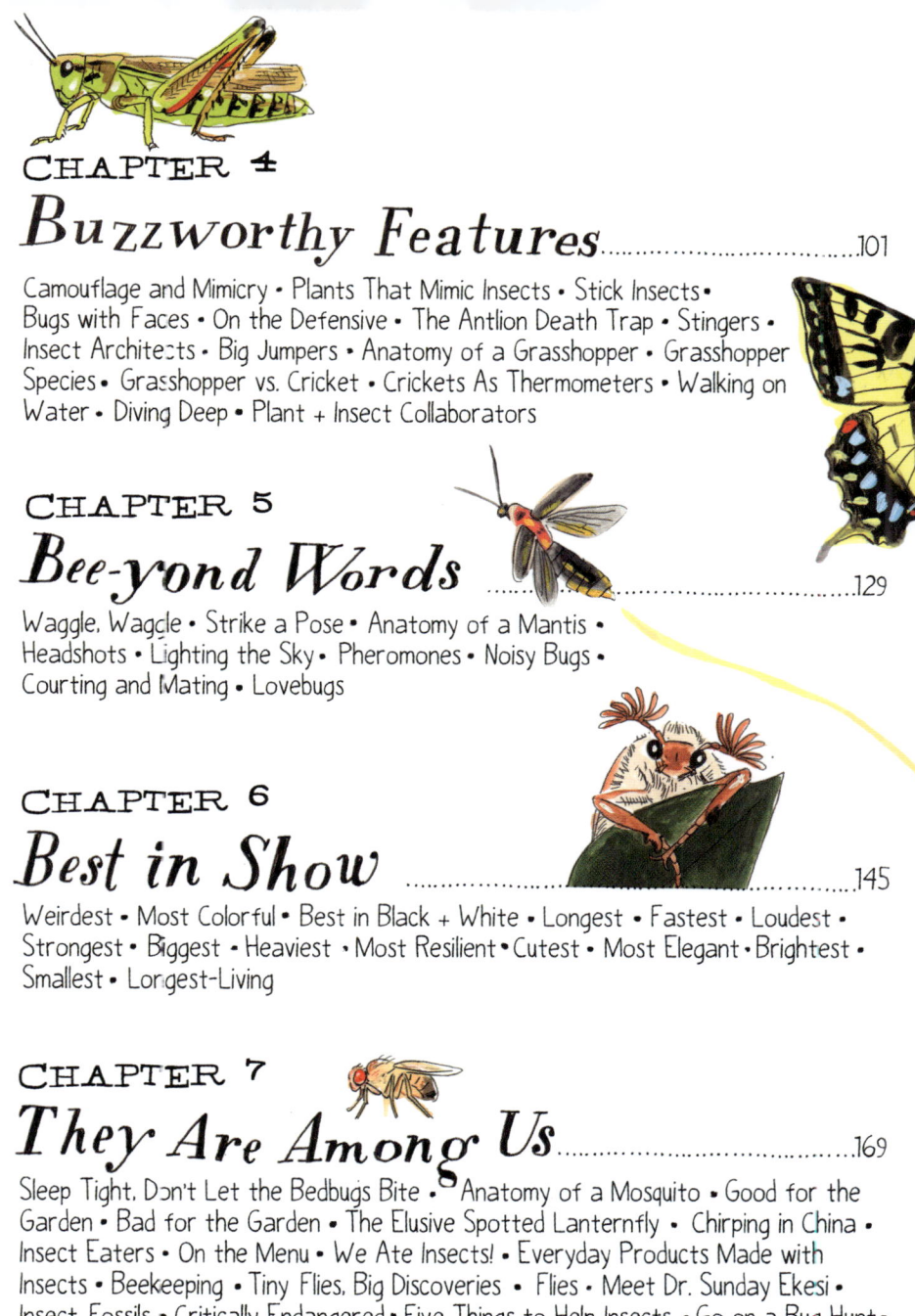

INTRODUCTION

As I worked on this book, memories of encounters with insects came rushing back, spanning from childhood moments to experiences as recent as last week. I started to write them all down:

Tracking the trails of ants marching across the countertop in my childhood home, trying to find where they were getting in—it drove my dad crazy.

Catching fireflies in my neighbor's yard and placing them in a glass jar with a lid pierced with holes, we would watch as they lit up together before releasing them. There were so many back then.

When my parents discovered that my sister and I had lice during a family trip to Disneyland, they spent hours in the hotel room shampooing our hair and painstakingly combing through it with a fine-tooth comb. By the end, there was a pile of lice and eggs on the floor that had to be swept away.

A monarch butterfly kept returning to our backyard, always landing on my mom. She wondered if it was a sign from her father, my granddad, who had recently passed away.

When my sister returned from Uganda with a large welt, we discovered it was a mango fly larva burrowed into her skin. My dad had to help her pop it out—probably the grossest thing I've ever witnessed.

I woke up with bites on my stomach and tore apart my Brooklyn apartment, convinced I had bed bugs. I even called a bed bug hunter who brought his dog to sniff every corner of my bedroom. After a thorough inspection, he reassured me that, like 90% of his clients, I was just a neurotic New Yorker.

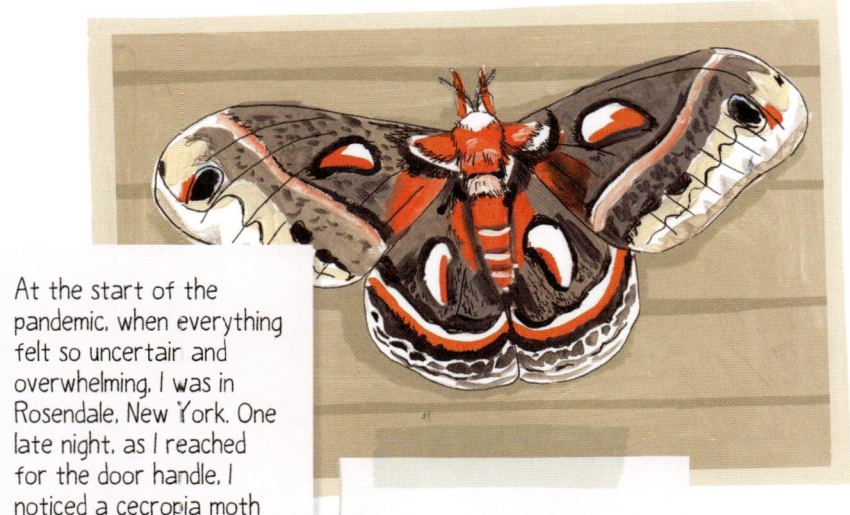

At the start of the pandemic, when everything felt so uncertain and overwhelming, I was in Rosendale, New York. One late night, as I reached for the door handle, I noticed a cecropia moth perched there. Its beauty was so striking and I watched it for a while. Somehow observing it made me feel that everything would be okay.

Going to see a giant wētā at the zoo in New Zealand and feeling the hairs on my arms stand up seeing its size.

I took a break from painting the final chapter of this book, and my husband and I spotted a large green caterpillar on the Brooklyn sidewalk. He challenged me to identify it, but embarrassingly, I couldn't, so I used an app to help. It turned out to be a polyphemus moth caterpillar. Using a piece of paper and a stick, we carefully moved it into the park.

We all have memories of past interactions with insects, but we sometimes overlook their tiny, buzzing lives in our day-to-day world. Where are they going? What are they eating? Why are they the colors they are? I loved learning about them! There were so many wonderful surprises, like discovering the perfect disguise of a dead leaf butterfly, or that the bagworm moth caterpillar builds a little log cabin on its back.

Now that I've painted ants up close, I pause when I see one in my kitchen. I observe its mandibles, the shape of its antennae, and the way its legs move. Creating this book reminded me to notice more, to stay curious, and to respect these remarkable creatures.

With this book, I invite you to remember and cherish your own insect memories.

xo Julia

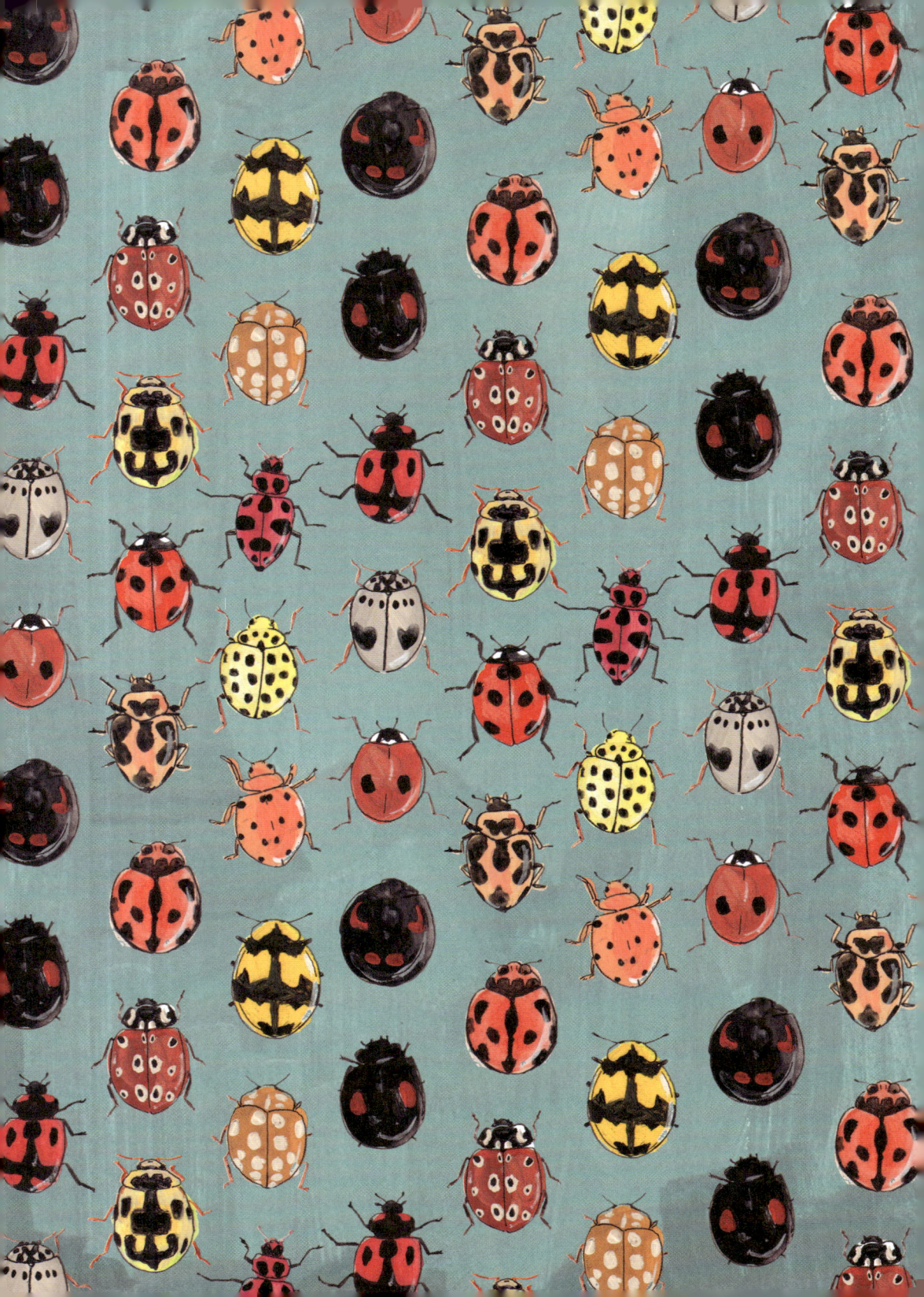

CHAPTER 1

Bugging Out

WHAT IS A BUG?

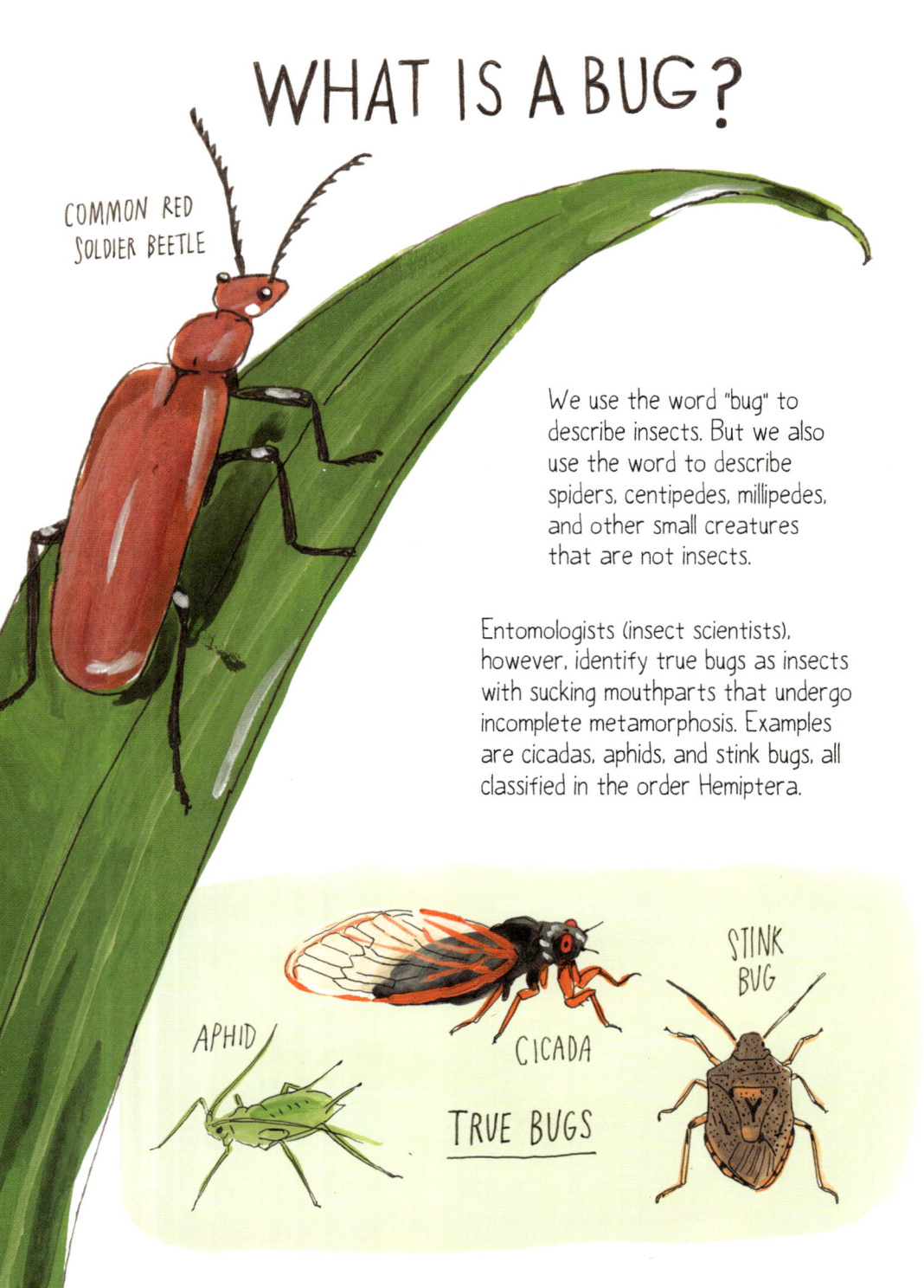

COMMON RED
SOLDIER BEETLE

We use the word "bug" to describe insects. But we also use the word to describe spiders, centipedes, millipedes, and other small creatures that are not insects.

Entomologists (insect scientists), however, identify true bugs as insects with sucking mouthparts that undergo incomplete metamorphosis. Examples are cicadas, aphids, and stink bugs, all classified in the order Hemiptera.

APHID

CICADA

STINK BUG

TRUE BUGS

INSECT VS. SPIDER

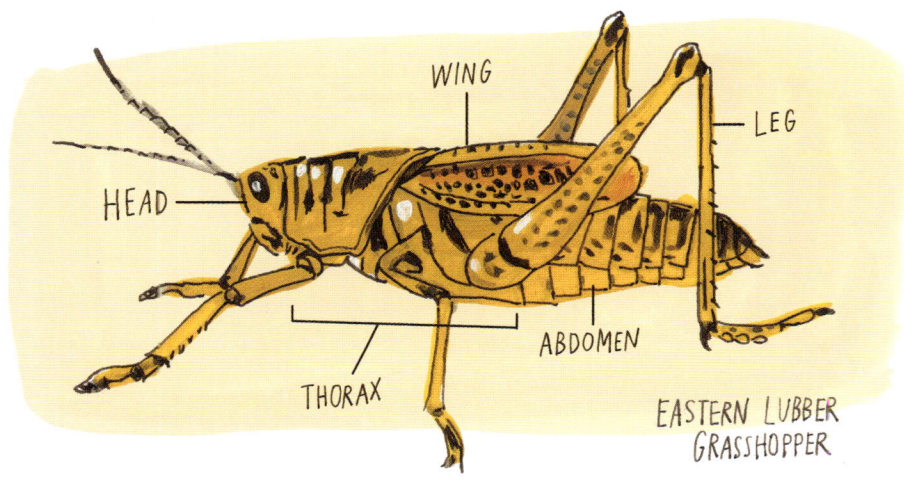

HEAD

WING

LEG

THORAX

ABDOMEN

EASTERN LUBBER
GRASSHOPPER

INSECTS (Insecta) have three body parts: head, thorax, and abdomen, as well as six legs, and typically wings.

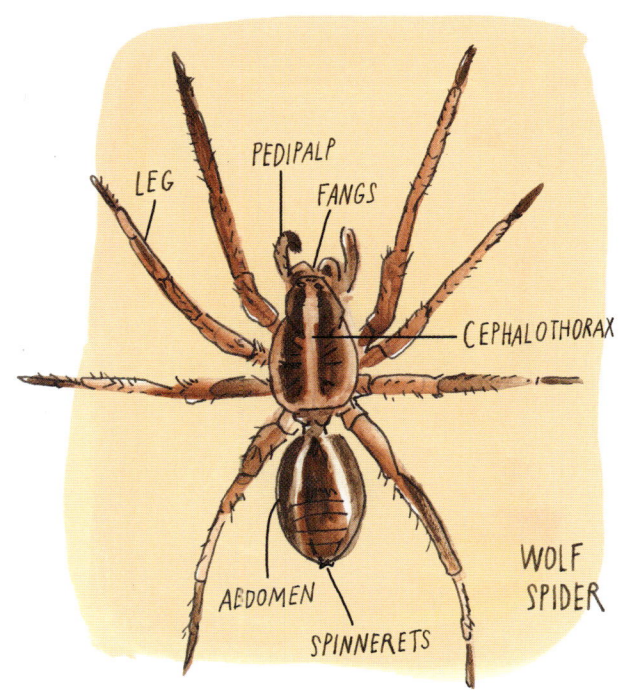

LEG

PEDIPALP

FANGS

CEPHALOTHORAX

ABDOMEN

SPINNERETS

WOLF
SPIDER

SPIDERS (Arachrida) have just two major body parts: cephalothorax and abdomen, plus fangs, pedipalps (feelers), and spinnerets for spinning threads for webs and cocoons. They also have eight legs.

Insects represent more than half of all living described species on our planet. With more than 1 million species of insects identified, scientists estimate that we have only discovered about 20 percent of the total number of species on Earth.

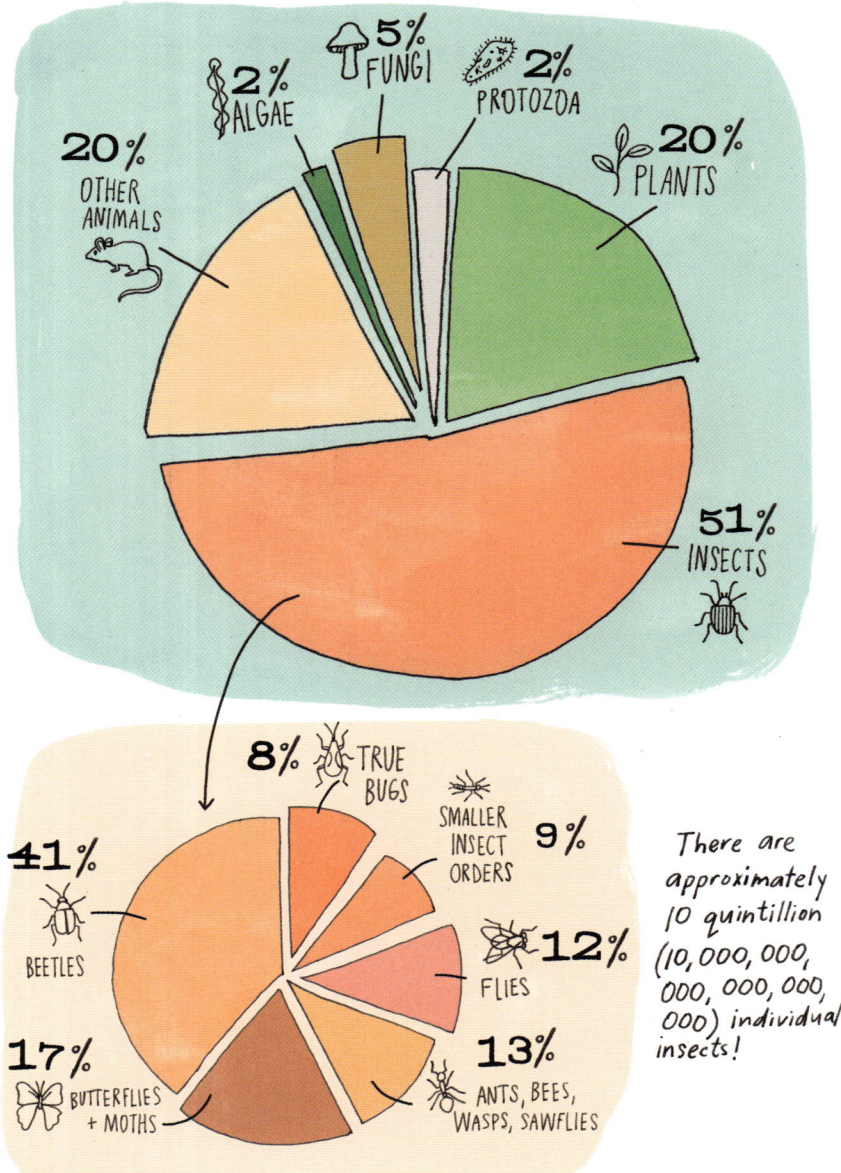

2% ALGAE

5% FUNGI

2% PROTOZOA

20% PLANTS

20% OTHER ANIMALS

51% INSECTS

8% TRUE BUGS

SMALLER INSECT ORDERS 9%

41% BEETLES

12% FLIES

17% BUTTERFLIES + MOTHS

13% ANTS, BEES, WASPS, SAWFLIES

There are approximately 10 quintillion (10,000,000, 000,000,000, 000) individual insects!

INSECT ORDERS

There are more than a million documented species of insects, most of which come from these seven orders within the family Insecta.

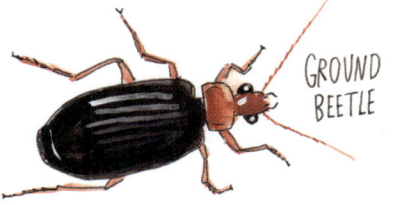

GROUND BEETLE

Coleoptera is beetles and is the largest order, comprising over 40 percent of all described insect species. They have strong shells and hard cases over their wings.

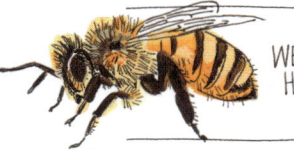

WESTERN HONEY BEE

Hymenoptera includes bees, wasps, hornets, and ants. Most species in this group have stingers.

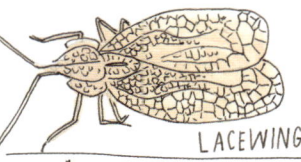

LACEWING

Hemiptera, or true bugs, have sharp, sucking mouthparts and usually colorful wings. Cicadas and aphids belong in this group.

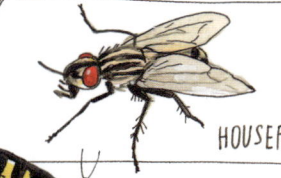

HOUSEFLY

Diptera contains flies, midges, and mosquitoes. They have a single pair of wings and bristly bodies.

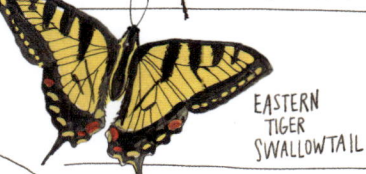

EASTERN TIGER SWALLOWTAIL

Lepidoptera is the order for butterflies and moths, which are characterized by having two pairs of wings covered in tiny scales.

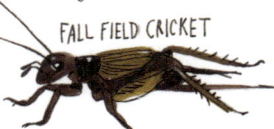

FALL FIELD CRICKET

Orthoptera have straight wings and long, powerful legs. This order includes crickets, grasshoppers, katydids, and locusts.

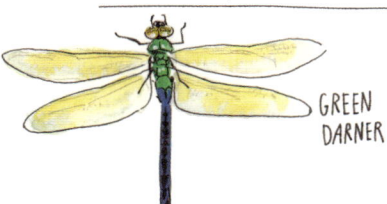

GREEN DARNER

Odonata are dragonflies and damselflies, which have long slim bodies, transparent wings, and large eyes.

WHAT IS AN ARTHROPOD?

Arthropods are invertebrate animals with segmented bodies, jointed limbs, and an exoskeleton. All insects are arthropods. All spiders are arthropods. However, not all arthropods are spiders or insects! Some sea creatures, such as crabs, shrimp, and barnacles, are also arthropods.

More than 80 percent of all living creatures are arthropods, the vast majority of which are insects. Arthropods can be divided into five branches.

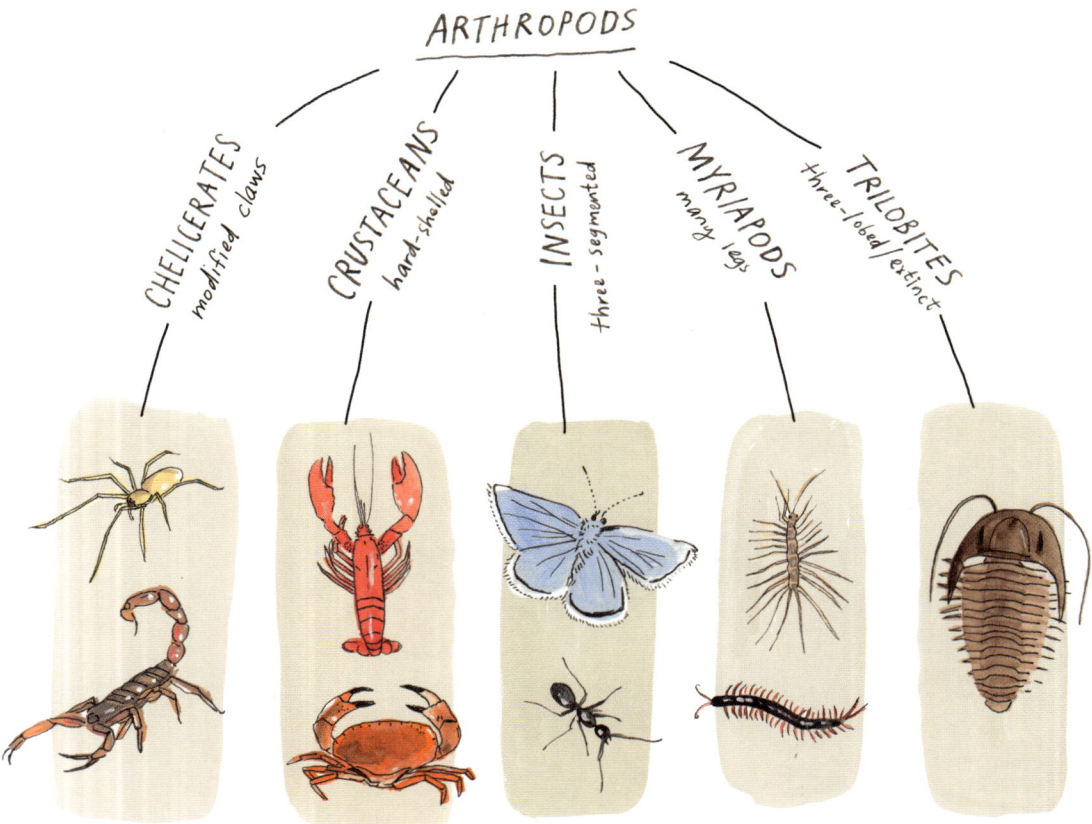

ARTHROPODS

CHELICERATES
modified claws

CRUSTACEANS
hard-shelled

INSECTS
three - segmented

MYRIAPODS
many legs

TRILOBITES
three-lobed/extinct

THE EXOSKELETON

Exoskeletons are a key feature among arthropods. Like a suit of armor, the exoskeleton protects the insect and keeps it from drying out.

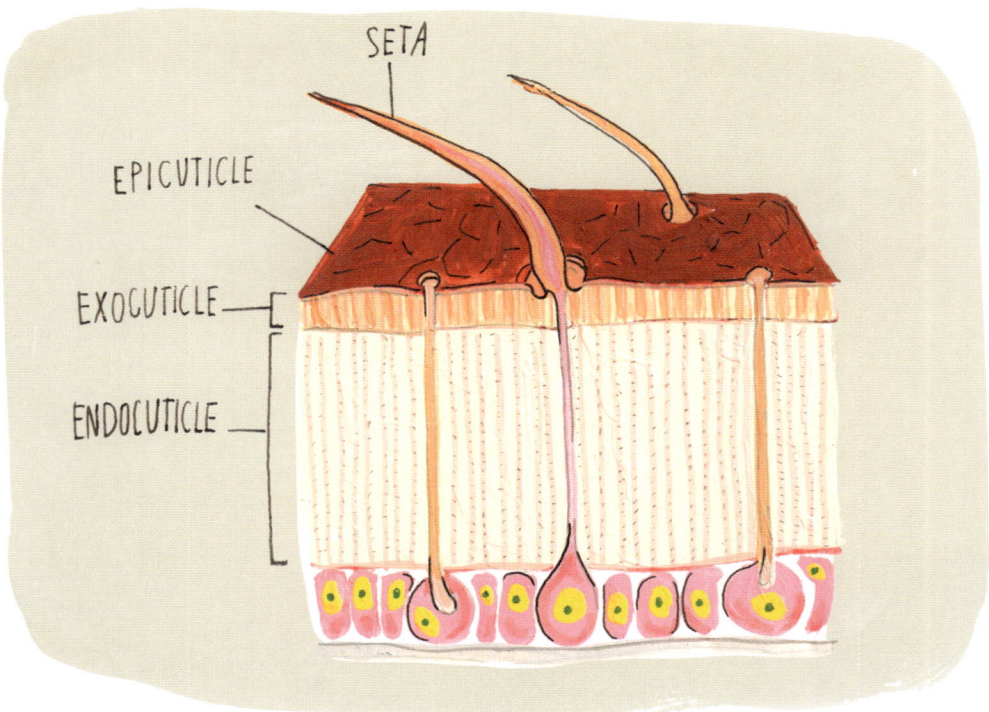

An insect's exoskeleton is made of three layers: the endocuticle, exocuticle, and epicuticle. The endocuticle and exocuticle are made of chitin and also sometimes calcium carbonate. The outermost layer, the epicuticle, contains a waxy protective substance. Although exoskeletons are rigid and durable, they have joints that allow the animals to move.

THESE CREATURES ARE <u>NOT</u> INSECTS

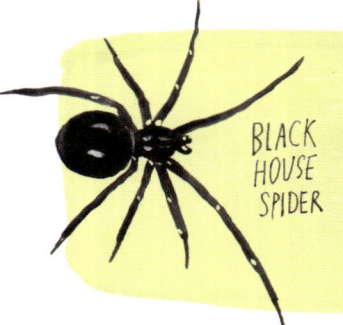

Spiders

BLACK HOUSE SPIDER

There are more than 45,000 species of spiders. All are air-breathing arthropods. Unlike insects, spiders have a fused head and thorax. They also do not have wings and antennae like insects.

Ticks

Ticks are often mistaken for insects, but like spiders they have two body parts and eight legs and no antennae.

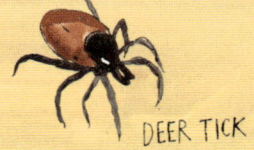

DEER TICK

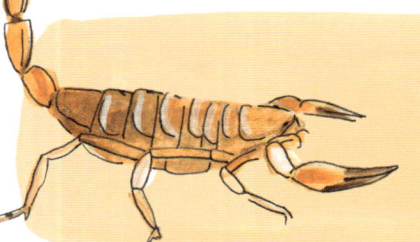

Scorpions

Cousins to spiders, mites, and ticks, these stinging, eight-legged arthropods are also part of the class Arachnida.

Worms

"Worms" refers to many different creatures that have a long, tubular body with no legs, and often lack eyes.

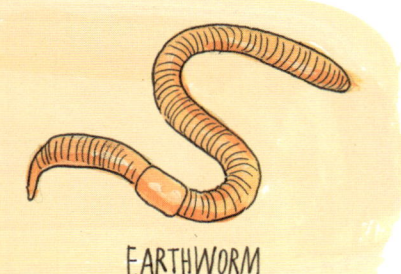

EARTHWORM

Pill bugs

Commonly referred to as "rollie pollies," pill bugs are in the crustacean group, placing them in closer relationship to shrimp and lobster than insects. Pill bugs are typically found in damp locations, under rocks or in rotting vegetation. Like armadillos (to which they are not related), they are known for their ability to roll into a ball when feeling threatened.

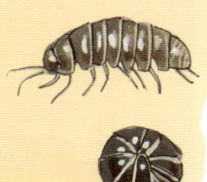

Slugs + Snails

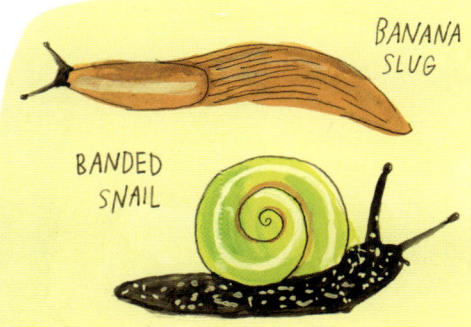

BANANA SLUG

BANDED SNAIL

Slugs and snails are molluscs (Mollusca), the same phylum as squid, cuttlefish, and octopuses; as well as mussels, clams, and oysters. Although slugs and snails are closely related, they are different species, and can be found throughout the world living on land and in water.

Centipedes + Millipedes

Centipedes and millipedes belong to the subphylum Myriapoda, which means "many feet." Despite common belief, centipedes don't have 100 legs. The number of legs varies from as few as 30 to nearly 400, but it's always an odd number of pairs.

Millipedes (Diplopoda) have two sets of legs per segment positioned directly under their bodies. And while the word "milli" comes from Latin for 1,000, it is rare for a millipede to have more than 750 legs.

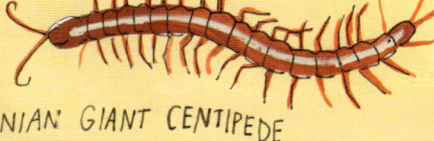

AMAZONIAN GIANT CENTIPEDE

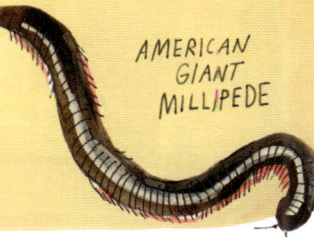

AMERICAN GIANT MILLIPEDE

A TIMELINE OF INSECTS

Recent studies suggest that the earliest insects lived more than 400 million years ago. This chart shows when many orders first appeared.

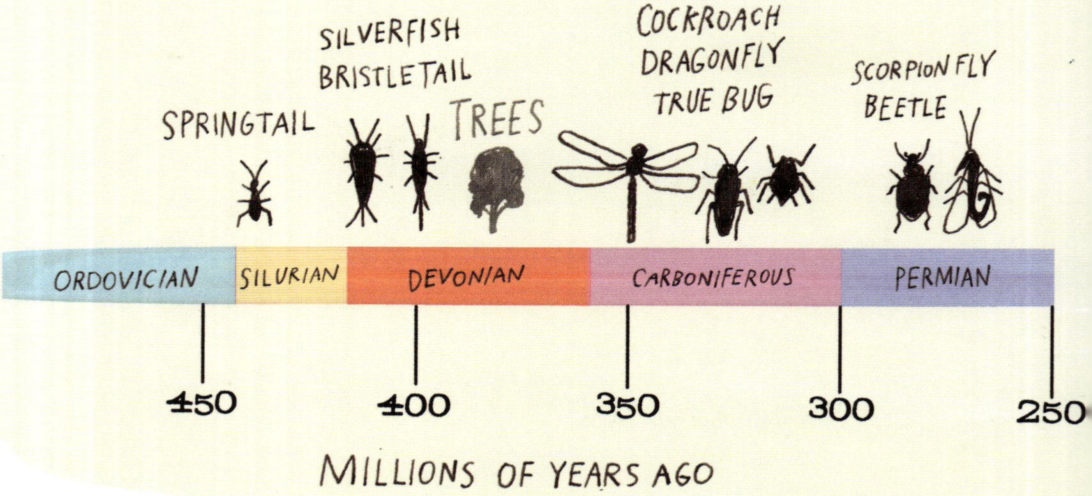

SPRINGTAIL

SILVERFISH
BRISTLETAIL

TREES

COCKROACH
DRAGONFLY
TRUE BUG

SCORPION FLY
BEETLE

| ORDOVICIAN | SILURIAN | DEVONIAN | CARBONIFEROUS | PERMIAN |

450   400   350   300   250

MILLIONS OF YEARS AGO

Escaping Extinction

Two of the most resilient insects are cockroaches and ants. Both can handle extreme conditions that few other creatures can survive.

Cockroaches can endure high levels of radiation, and when exposed to various toxins, they quickly build up a resistance. In addition, they can eat almost anything, and can go long periods without food and water.

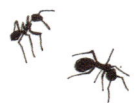

Insects have survived multiple mass extinction events throughout their evolutionary history. This is due to their small size, high reproduction rates, and ability to adapt.

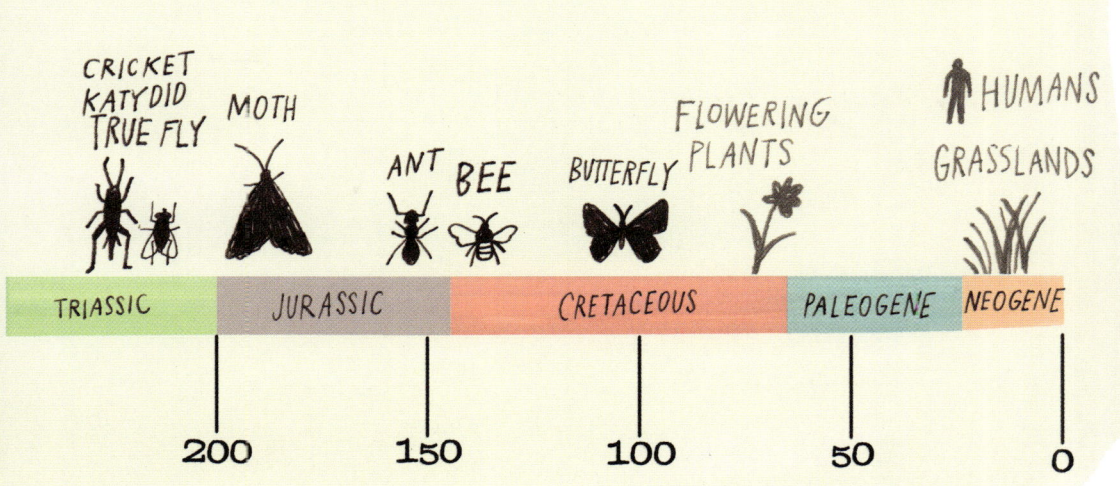

CRICKET
KATYDID
TRUE FLY   MOTH   ANT BEE   BUTTERFLY FLOWERING PLANTS   HUMANS
                                                                          GRASSLANDS

| TRIASSIC | JURASSIC | CRETACEOUS | PALEOGENE | NEOGENE |

200   150   100   50   0

More than 12,000 species of ants live in a wide variety of environments and climates. These insects are well adapted to handle severe weather events. Additionally, in the case of pandemic, the colony is quick to sacrifice members that are infected, helping to stop the spread of disease.

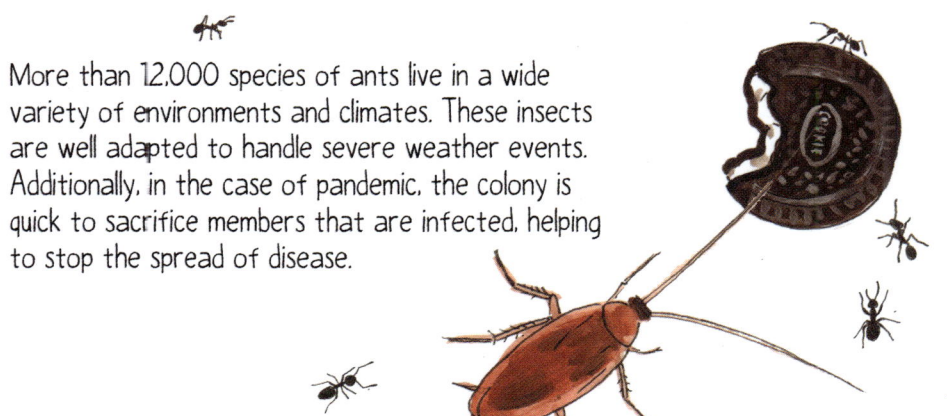

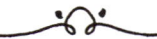

ANATOMY OF AN INSECT

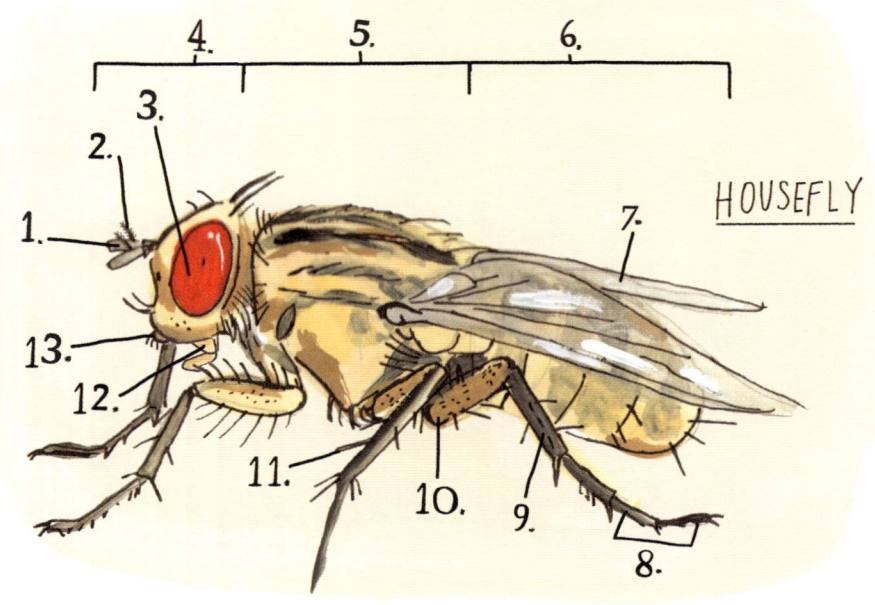

HOUSEFLY

1. antenna
2. arista
3. compound eye
4. head
5. thorax
6. abdomen
7. wing
8. tarsus
9. tibia
10. femur
11. spur
12. maxillary palps
13. labium

The head contains the eyes, antennae, and mouthparts. It may also feature sensory organs such as the arista and palps. The thorax is the middle body part, to which the insect's wings and legs are connected. The abdomen, the largest body part, contains most of the digestive and reproductive organs, as well as most of the spiracles (openings) for breathing.

INSIDE AN INSECT

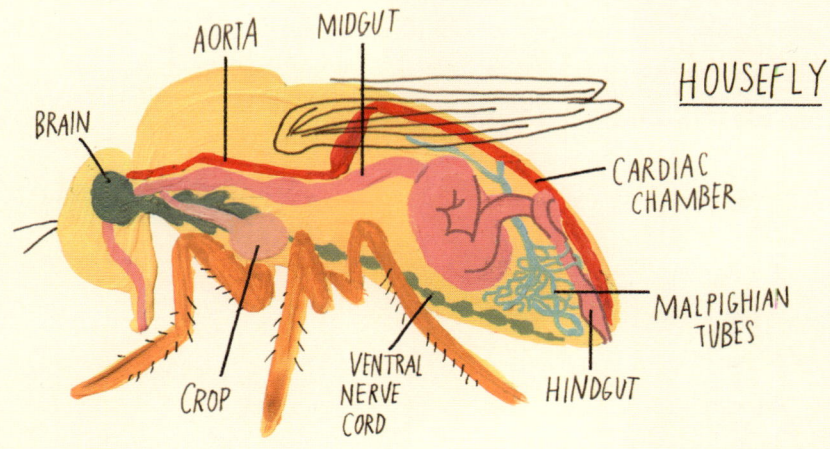

AORTA

MIDGUT

HOUSEFLY

BRAIN

CARDIAC
CHAMBER

MALPIGHIAN
TUBES

CROP

VENTRAL
NERVE
CORD

HINDGUT

The insect's digestive system is divided into three sections that break down food and absorb essential nutrients: the foregut, which includes the crop; midgut; and hindgut. A single blood vessel, the aorta, functions as a pump, directing blood flow throughout the body. The ventral nerve cord is the main pathway for sending nerve signals between the brain and the rest of the body.

HOW INSECTS SMELL

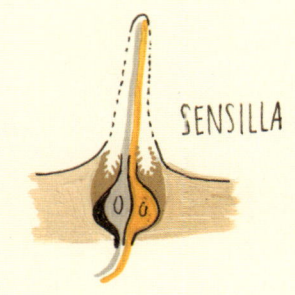

SENSILLA

Unlike humans, insects possess several sense organs, called sensilla, mostly found in the antennae. In some species, additional sensilla may be located on the mouthparts, wings, or even the genitalia. Odorants penetrate the sensilla through pores, where they meet olfactory receptor neurons that process and identify the odors.

WINGS AND THINGS

Within the animal kingdom, insects are the only invertebrates that can fly. Each insect wing consists of a membrane supported by a series of structural veins. This pattern of veins, along with wing shape and size, is helpful in identifying insects. While not all insects can fly, most have one or two sets of wings, referred to as the forewings and the hindwings.

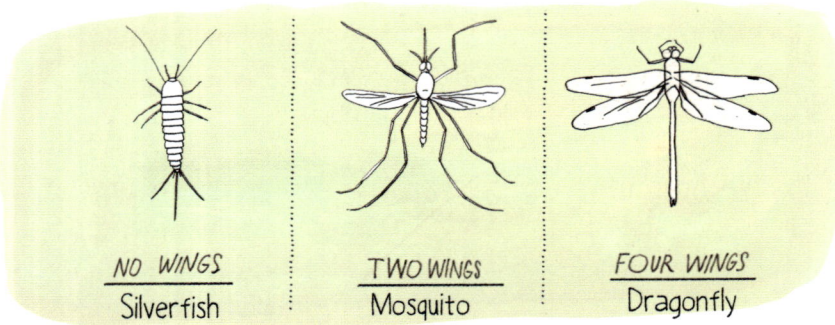

| NO WINGS | TWO WINGS | FOUR WINGS |
| Silverfish | Mosquito | Dragonfly |

There are many variations in flight patterns among different species of insects. While beetles may appear rather clumsy during flight, the graceful maneuvers of dragonflies can be sublime. Erratic flight patterns are generally attributed to survival—the escape from predators.

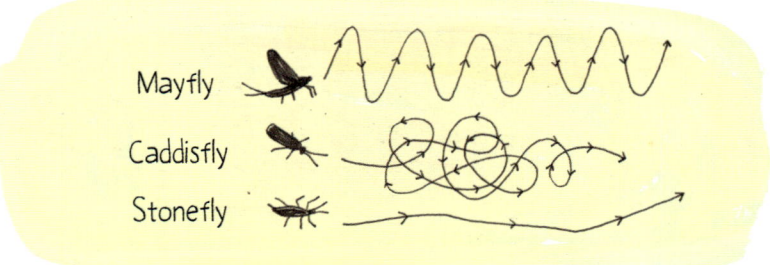

Mayfly

Caddisfly

Stonefly

Most insects use a method of wing spiraling to create lift. The front edge of the wing dips down and forward and then up and backward, outlining a horizontal figure eight.

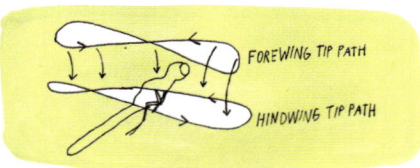

FOREWING TIP PATH

HINDWING TIP PATH

SHAPES OF INSECT WINGS

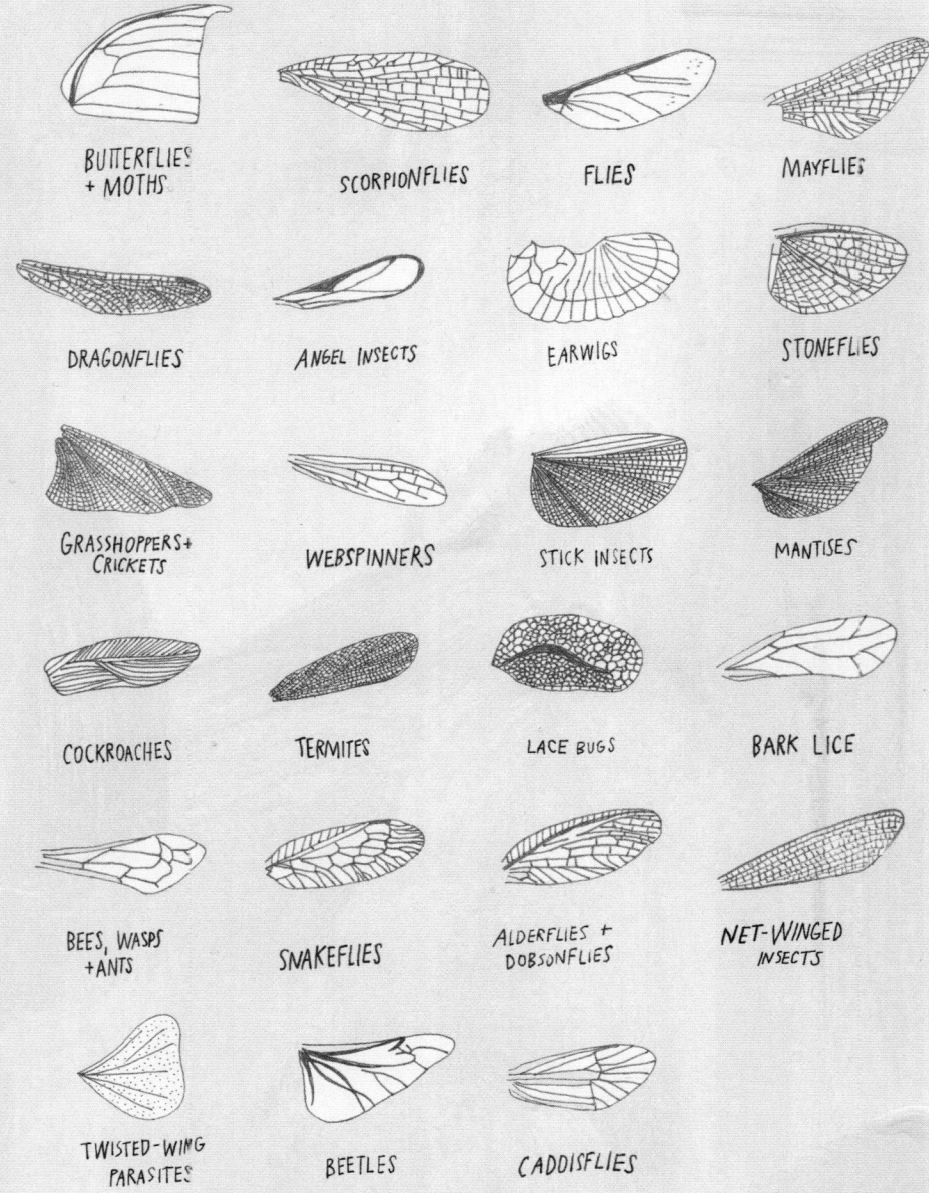

BUTTERFLIES + MOTHS

SCORPIONFLIES

FLIES

MAYFLIES

DRAGONFLIES

ANGEL INSECTS

EARWIGS

STONEFLIES

GRASSHOPPERS + CRICKETS

WEBSPINNERS

STICK INSECTS

MANTISES

COCKROACHES

TERMITES

LACE BUGS

BARK LICE

BEES, WASPS + ANTS

SNAKEFLIES

ALDERFLIES + DOBSONFLIES

NET-WINGED INSECTS

TWISTED-WING PARASITES

BEETLES

CADDISFLIES

KINDS OF WINGS

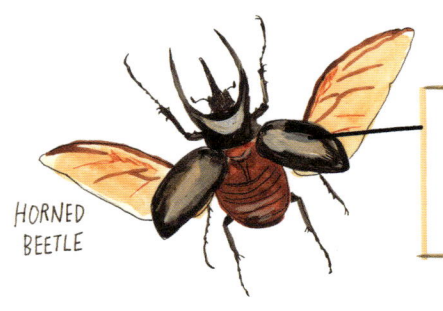

HORNED
BEETLE

Elytra

These are the hard-shell forewings of beetles that protect the hindwings.

Fissured

Found in the plume moth and a few other species, this design has a forked appearance.

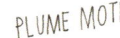

PLUME MOTH

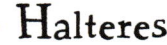

CRANE FLY

Halteres

These smaller, modified hindwings move at the same frequency as the larger wings, but in the opposite direction, which helps with stability.

Hemelytra

Found on true bugs such as the red cotton bug, these wings are leathery at the base and membranous at the tip.

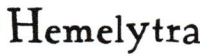

RED COTTON BUG

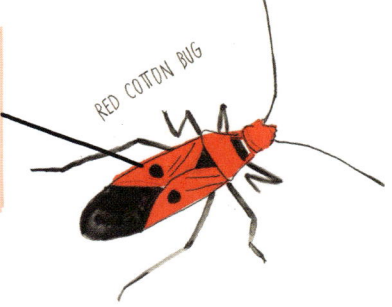

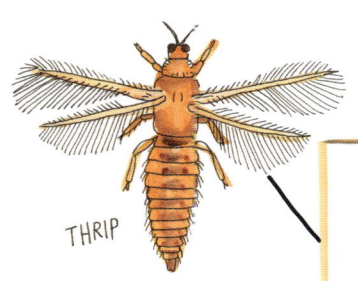

THRIP

Fringed

Found primarily in thrips, these conspicuous wings help the tiny insects generate the extra force needed for flight.

Membranous

These thin, transparent wings are found on dragonflies, honey bees, and termites.

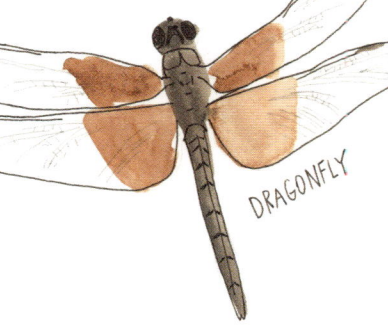

DRAGONFLY

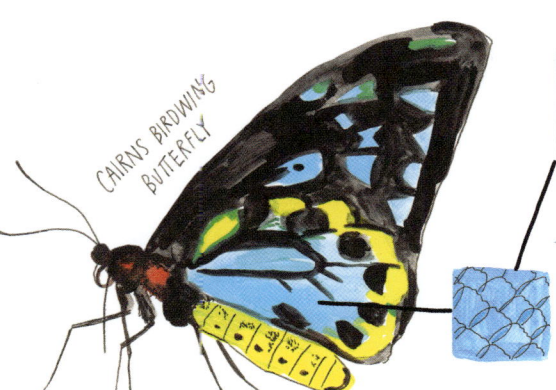

CAIRNS BIRDWING BUTTERFLY

Scaly

The word *lepidoptera* means "scaly wing." Moths and butterflies have thousands of tiny scales that overlap.

Tegmina

The thick forewings of such insects as grasshoppers and cockroaches protect the more vulnerable hindwings.

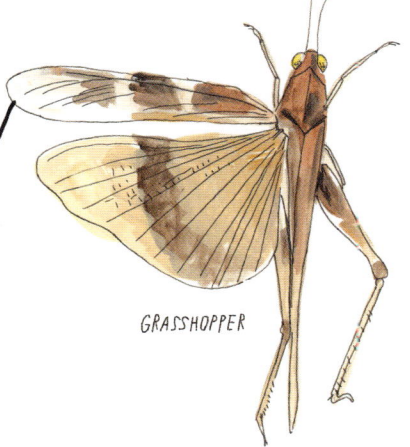

GRASSHOPPER

TYPES OF LEGS

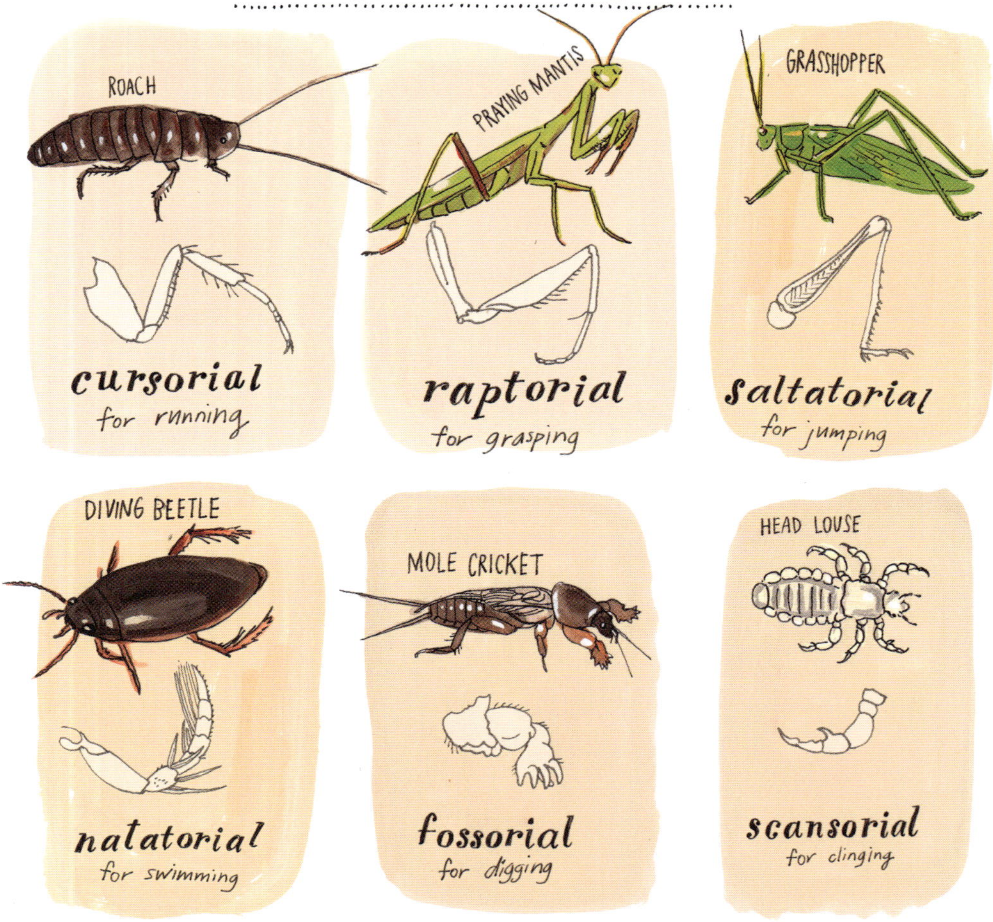

ROACH

cursorial
for running

PRAYING MANTIS

raptorial
for grasping

GRASSHOPPER

saltatorial
for jumping

DIVING BEETLE

natatorial
for swimming

MOLE CRICKET

fossorial
for digging

HEAD LOUSE

scansorial
for clinging

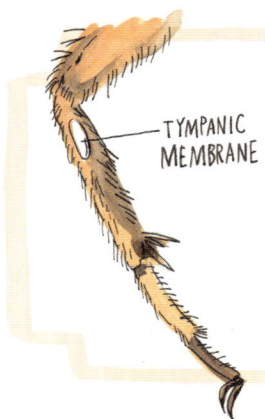

TYMPANIC MEMBRANE

Some insects, like crickets or grasshoppers, have something called "tympanic membranes" on their legs. These super-sensitive membranes can feel the vibrations made by sound waves.

INSECT EYES

Most insects have two types of eyes, simple and compound. The simple eyes (ocelli) are very small and made of a single lens. The compound eyes are the bulging eyes on each side of an insect's head consisting of many small lenses.

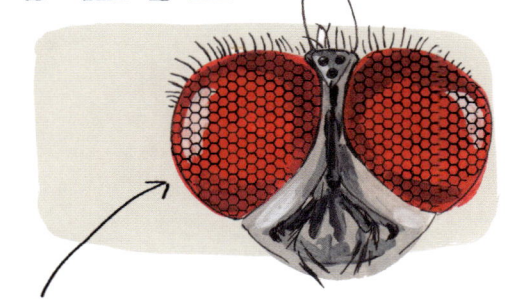

Flies have 4,000 lenses on their compound eyes. Bees can have up to 8,500.

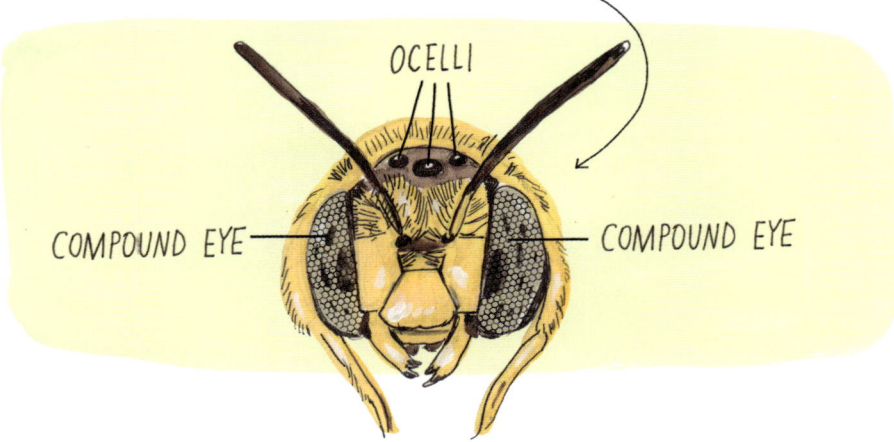

OCELLI

COMPOUND EYE —

— COMPOUND EYE

Simple eyes see light and shadow. Lenses in the compound eyes each have a slightly different view. The various images from the simple eyes and numerous compound eye lenses are processed in the brain, giving the insect a near 360-degree view, in color and with fine detail.

Dragonflies have 30,000! →

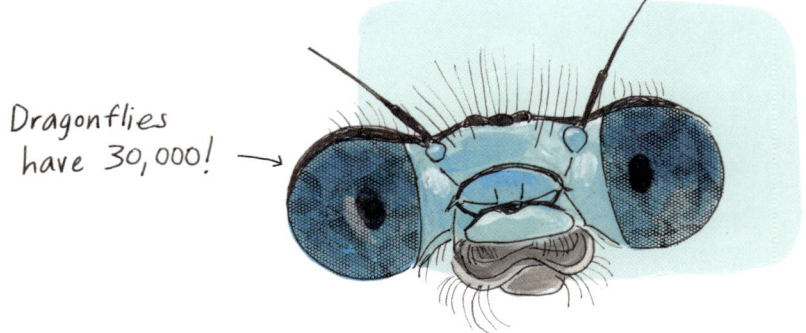

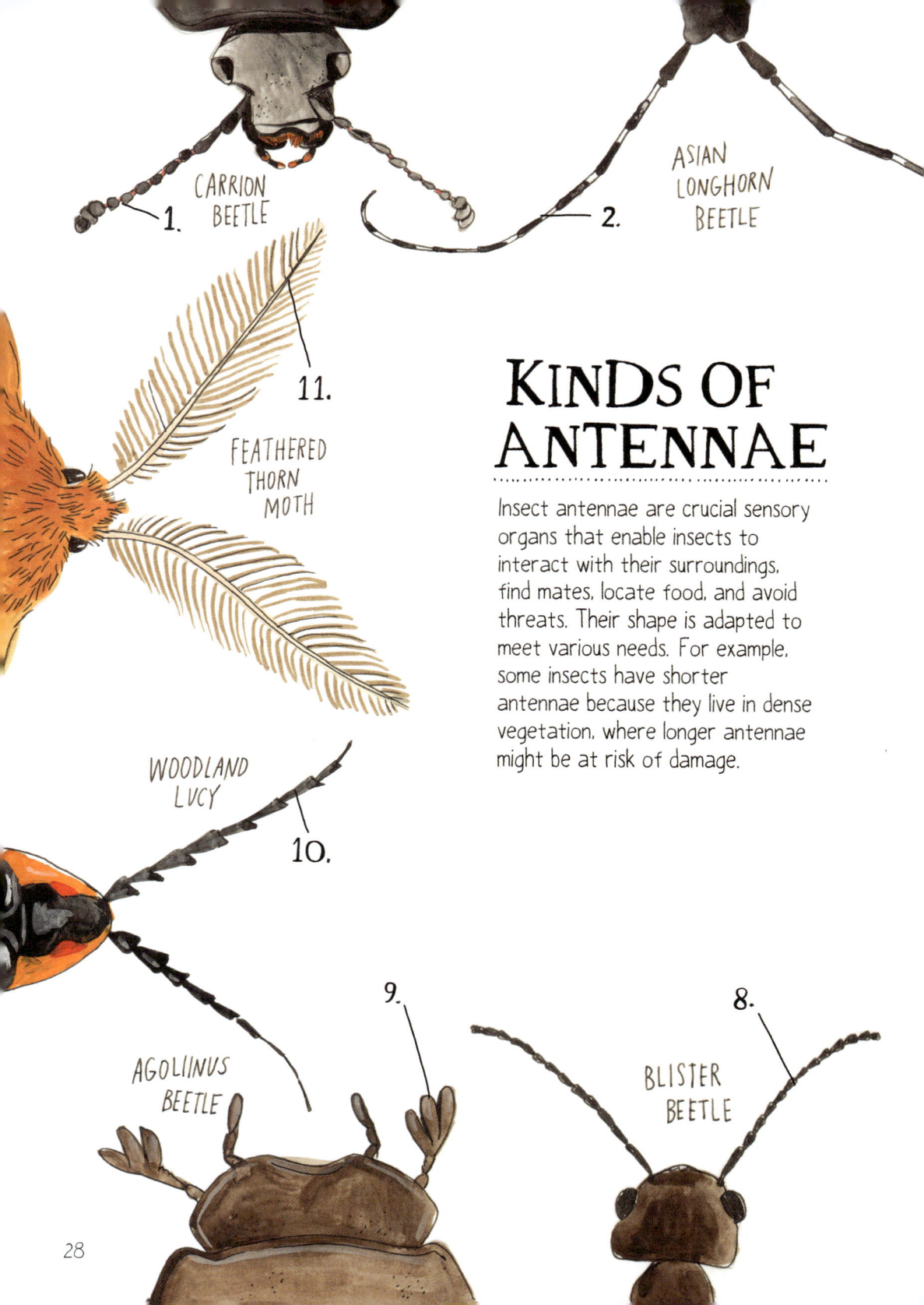

CARRION BEETLE 1.

ASIAN LONGHORN BEETLE 2.

11.

FEATHERED THORN MOTH

KINDS OF ANTENNAE

Insect antennae are crucial sensory organs that enable insects to interact with their surroundings, find mates, locate food, and avoid threats. Their shape is adapted to meet various needs. For example, some insects have shorter antennae because they live in dense vegetation, where longer antennae might be at risk of damage.

WOODLAND LUCY

10.

9.

8.

AGOLIINUS BEETLE

BLISTER BEETLE

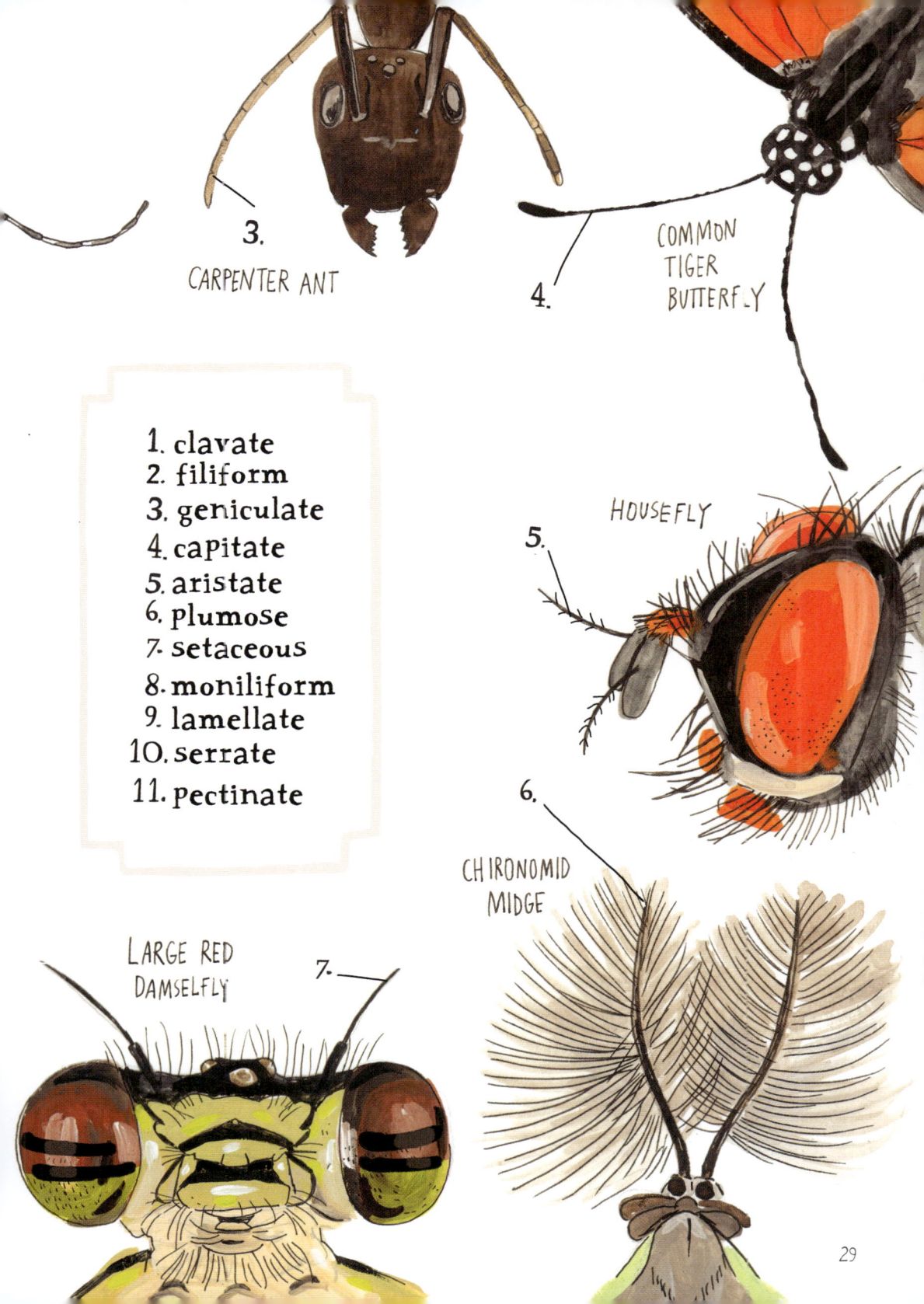

3.

CARPENTER ANT

4.

COMMON
TIGER
BUTTERFLY

1. clavate
2. filiform
3. geniculate
4. capitate
5. aristate
6. plumose
7. setaceous
8. moniliform
9. lamellate
10. serrate
11. pectinate

HOUSEFLY

5.

6.

CHIRONOMID
MIDGE

LARGE RED
DAMSELFLY

7.

WHAT DO INSECTS EAT? EVERYTHING!

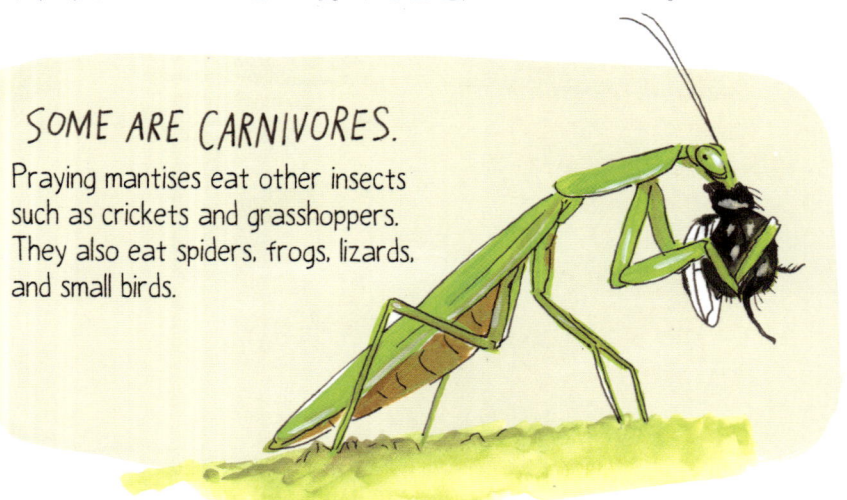

SOME ARE CARNIVORES.

Praying mantises eat other insects such as crickets and grasshoppers. They also eat spiders, frogs, lizards, and small birds.

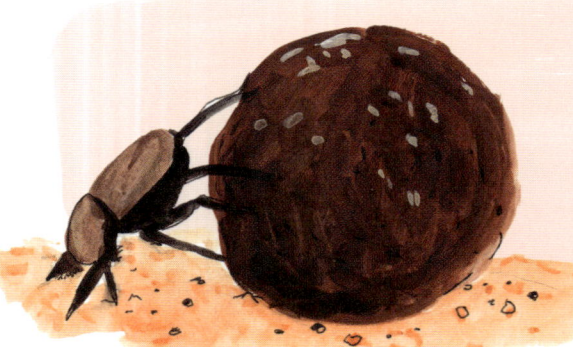

SOME EAT POOP.

Not only do dung beetles eat the dung of other animals, they also lay their eggs in it.

SOME LIKE SWEETS.

Bees eat nectar and pollen from plants. They gather the nectar and pollen to feed the larvae within their colony.

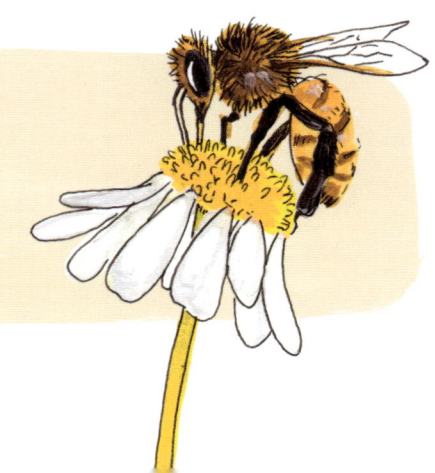

SOME EAT TOXINS TO WARD OFF PREDATORS.

Monarch caterpillars and milkweed bugs eat only milkweed, which is highly toxic to many insects, as well as chickens, cattle, sheep, dogs, and even humans.

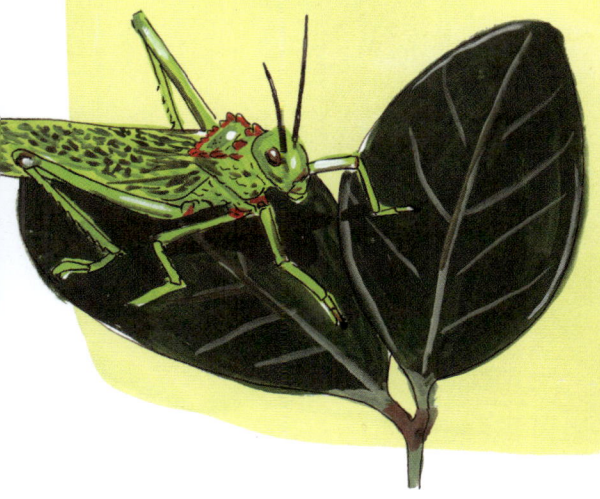

The African bush grasshopper feeds on plants such as the poison arrow tree (*Acokanthera oppositifolia*), which, as the name suggests, contains a sap that Indigenous people in Africa use to hunt with.

SOME EAT OUR FOOD.

Farmers spend millions on pesticides to control potato beetles, corn rootworms, and Khapra beetles that feed on crops and grains.

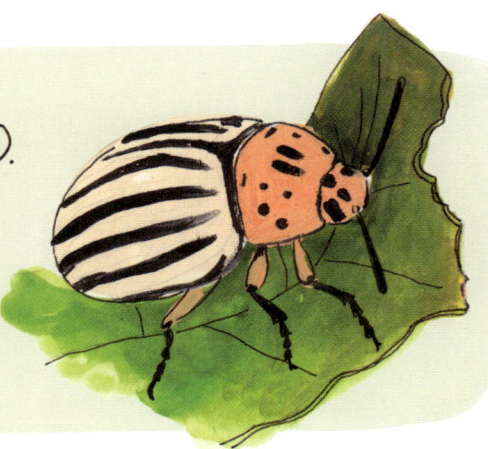

HOW INSECTS EAT

CHEWING

With their saw-shaped mandibles, grasshoppers can bite into leaves. The maxilla tastes the food, and the labium and labrum push it into the mouth.

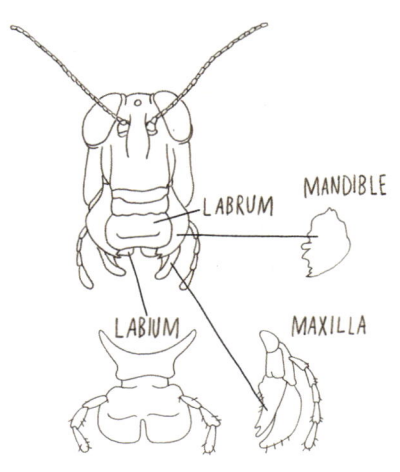

LABRUM

MANDIBLE

LABIUM

MAXILLA

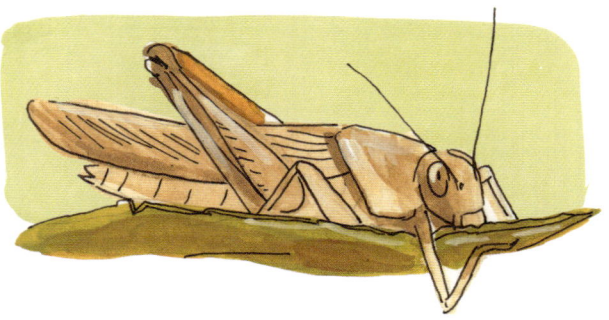

LAPPING

Flies use their proboscis to squirt a digestive fluid into their food. The fluid breaks the substance into mush that can be licked up.

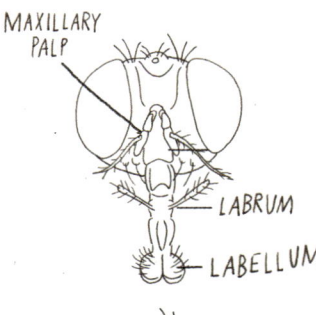

MAXILLARY PALP

LABRUM

LABELLUM

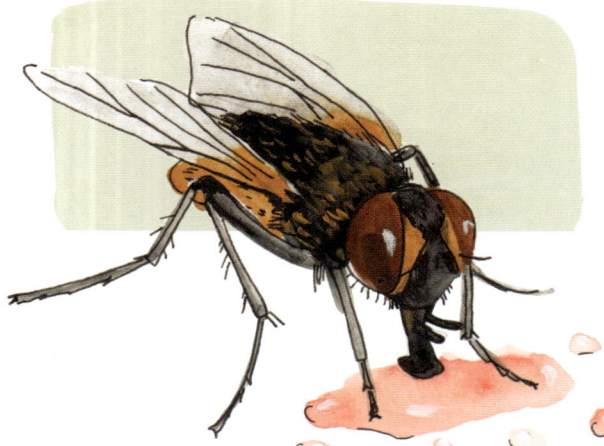

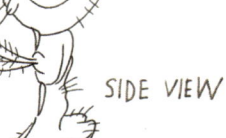

SIDE VIEW

PIERCING

Mosquitoes have needle-like mandibles that hold the skin apart while the maxillae pierce it. Like the fly, they pump digestive fluids down before sucking the animal's juices back up.

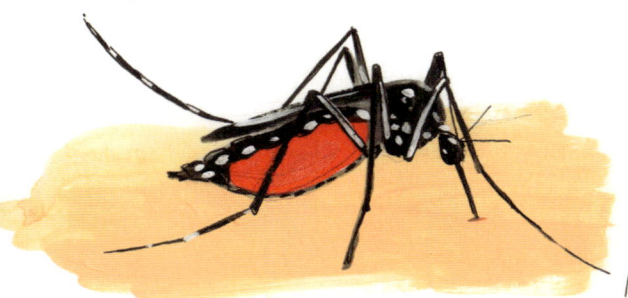

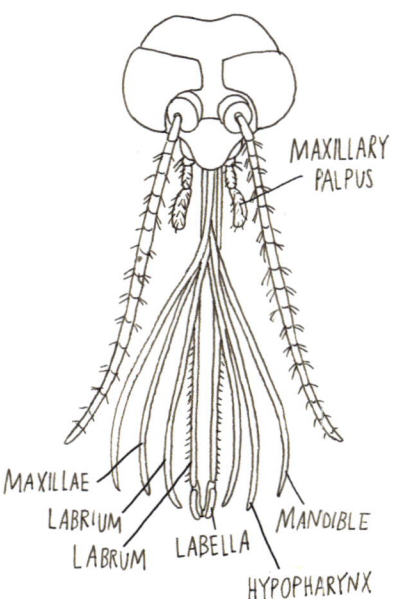

MAXILLARY PALPUS

MAXILLAE
LABRIUM
LABRUM
LABELLA
MANDIBLE
HYPOPHARYNX

SUCKING

The mouthparts of butterflies and moths are long, straw-like tubes for sucking nectar from flowers. Between sips, the proboscis is kept rolled up.

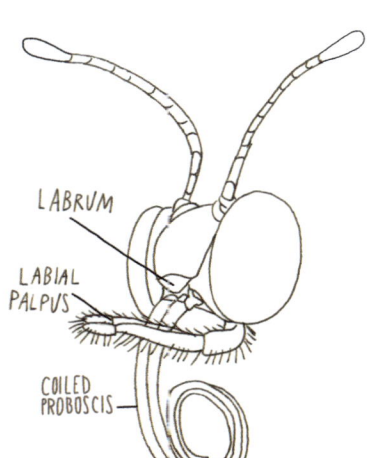

LABRUM

LABIAL PALPUS

COILED PROBOSCIS

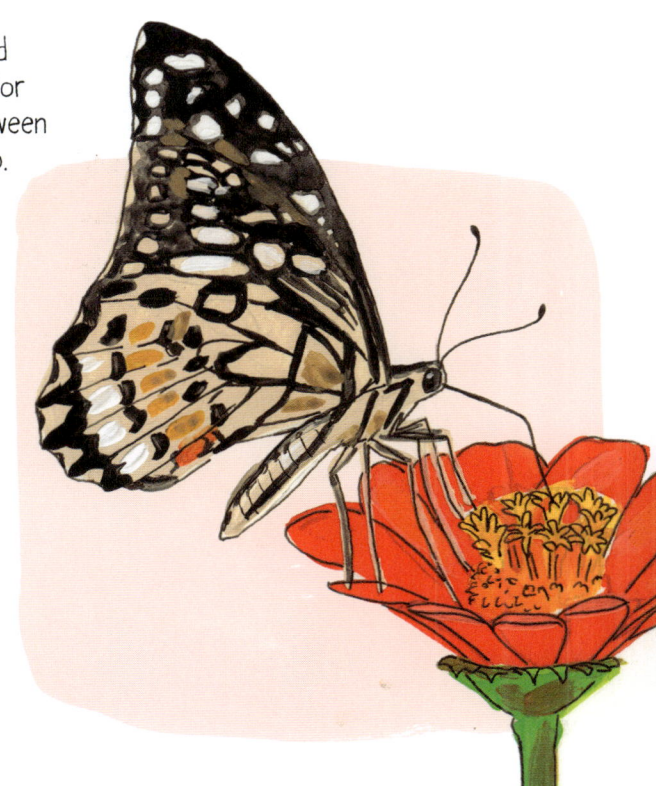

INSECT HABITATS

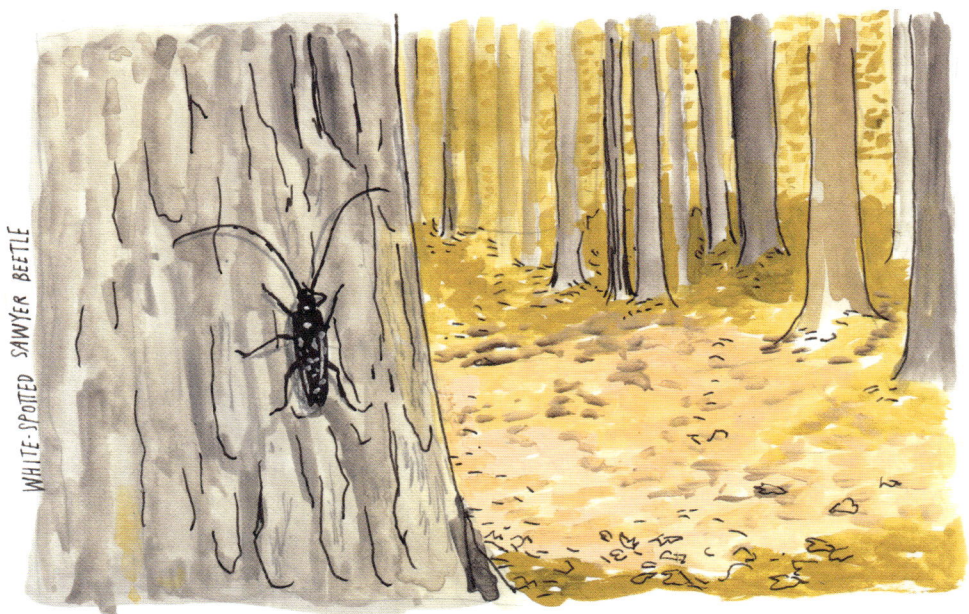

WHITE-SPOTED SAWYER BEETLE

FORESTS

The forest is home to many insects, which reside on trees, in leaf piles, and in the shaded undergrowth, including tree ants, wood-boring beetles, forest-dwelling butterflies, and woodland ground beetles.

GRASSLANDS

Insects can be found in grassy areas, living among the blades of grass or within the soil. They might also inhabit areas with shrubs and bushes. Grasshoppers, bees, butterflies, ants, and ladybugs are a few common insects found in grasslands.

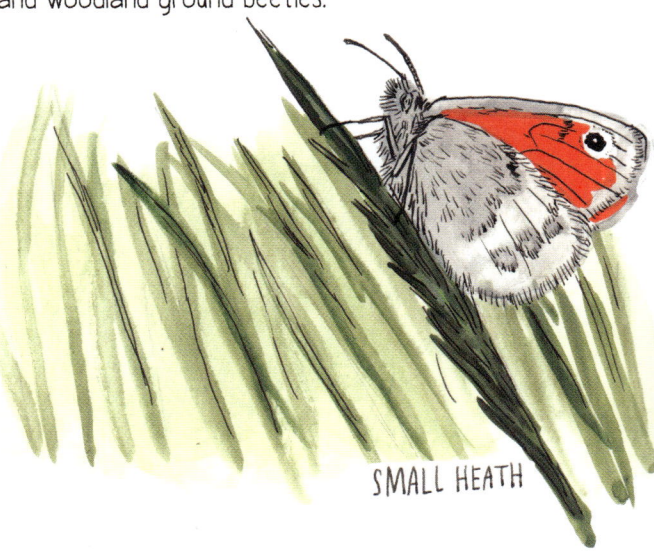

SMALL HEATH

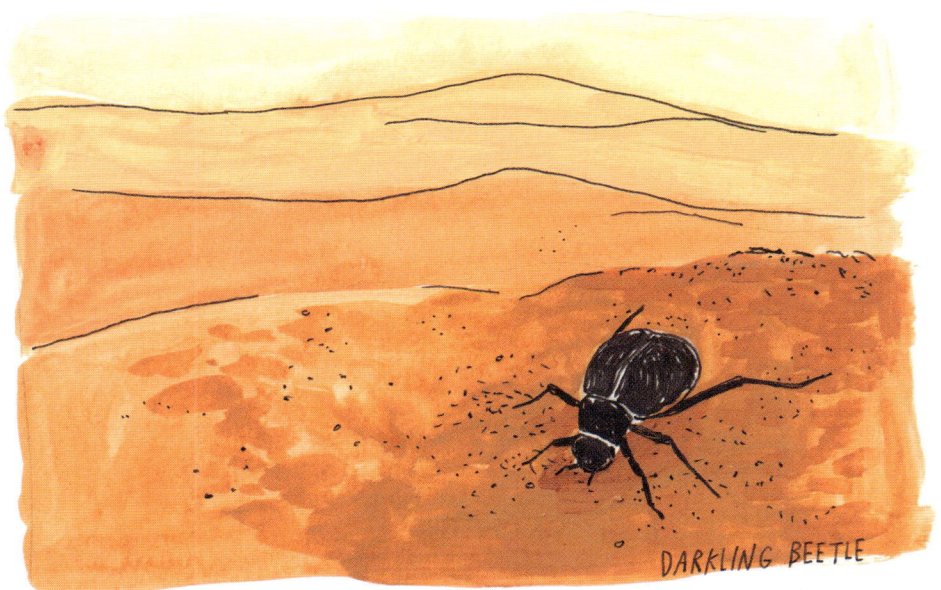

DARKLING BEETLE

DESERTS

Insects may burrow in the sand or find refuge under rocks where it's cooler and less exposed to the sun. Examples include desert ants, desert dung beetles, desert locusts, darkling beetles, Jerusalem crickets, and sand wasps.

WATER + WETLANDS

Many insects live near, on, and in bodies of water. Wetlands provide a diverse and rich habitat for various insects due to their unique mix of water and land. Water striders, dragonflies, mosquitoes, mayflies, and aquatic beetles thrive in this environment.

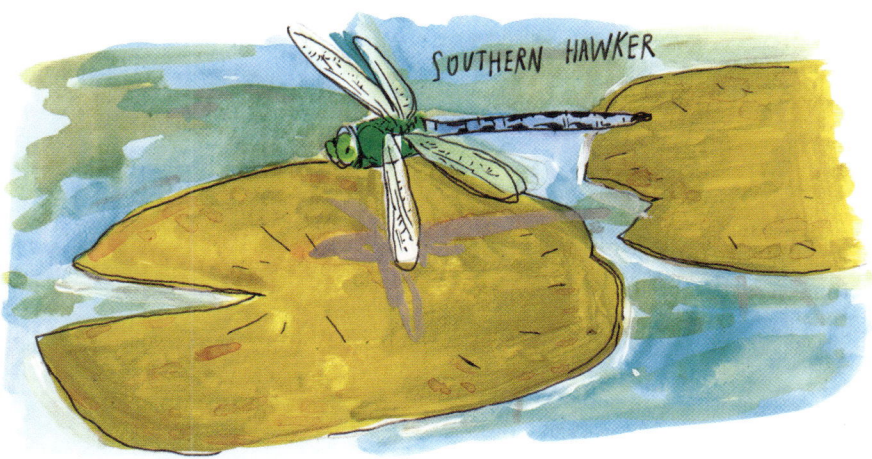

SOUTHERN HAWKER

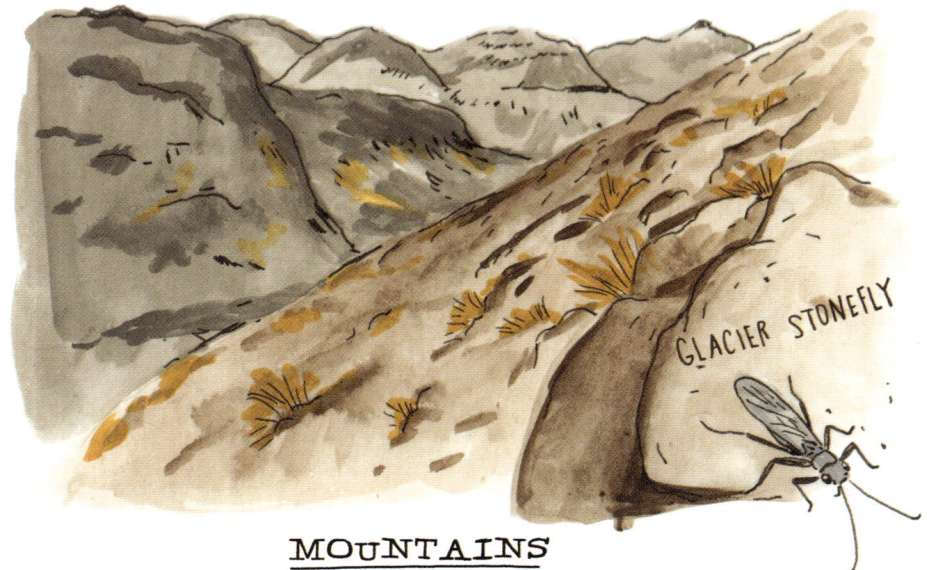

GLACIER STONEFLY

MOUNTAINS

Insects are found in mountainous regions, adapting to the specific conditions at different altitudes, living on rocky surfaces, or among high-altitude vegetation. Mountain stoneflies, alpine beetles, and high-altitude flies all play a role in this ecosystem.

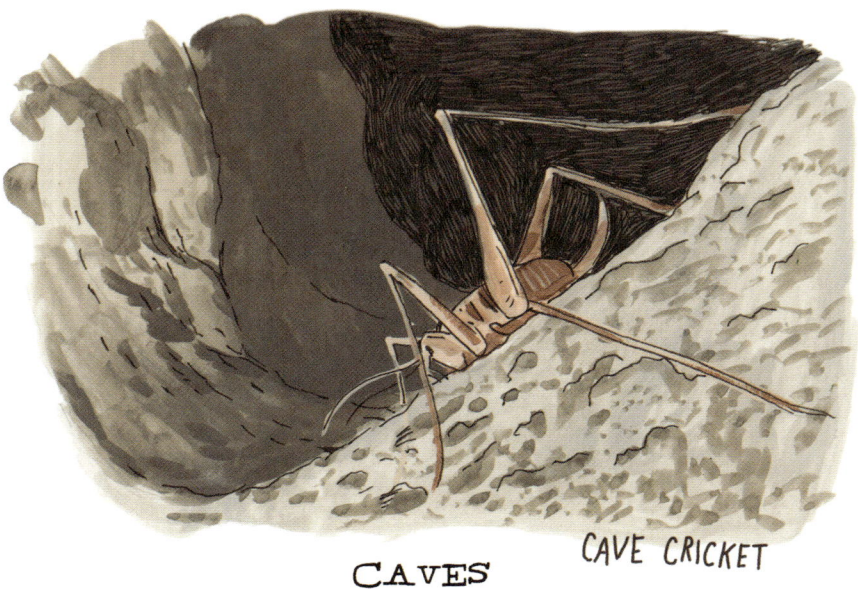

CAVE CRICKET

CAVES

Some insects live in caves, where they're protected from the elements. Cave crickets, stoneflies, springtails, and cave beetles have all adapted to survive constant darkness, high humidity, and limited food sources.

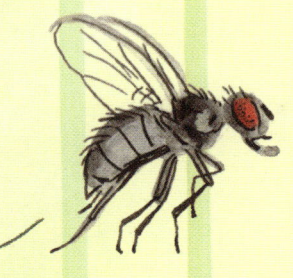

URBAN AREAS

Of course, insects also thrive in cities and towns, finding food and shelter created by humans. They reside in gardens, parks, and even inside our homes, finding shelter in cracks and crevices, or around plants. No doubt you are familiar with houseflies, cockroaches, ants, mosquitoes, and moths, just to name a few.

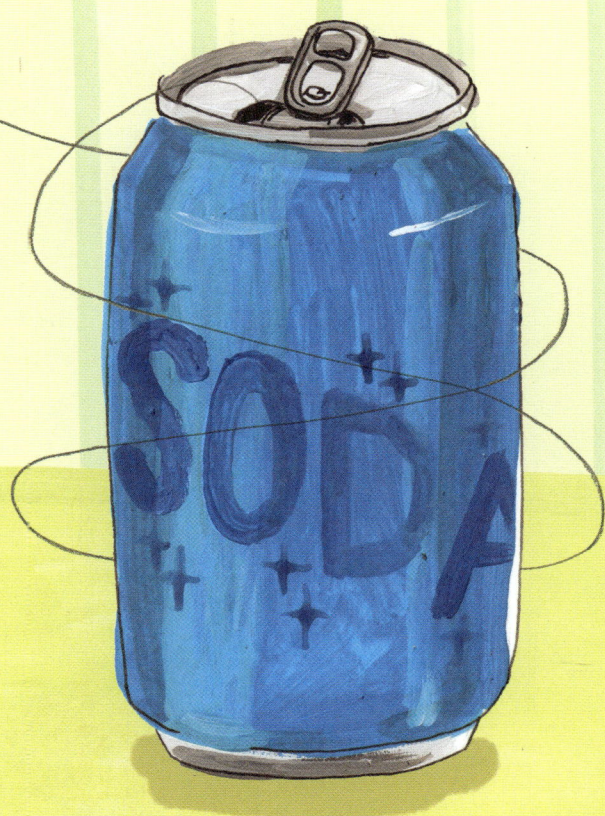

JAPANESE BEETLE

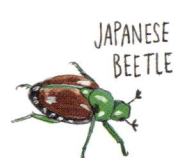

CARRION BEETLE

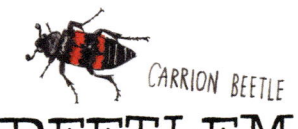

GREEN JUNE BEETLE

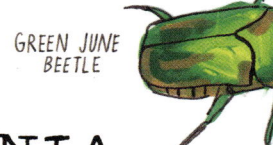

BEETLEMANIA

There are more than 400,000 known species of beetles, making them the largest group of insects on Earth. The order Coleoptera is the largest order in the entire animal kingdom.

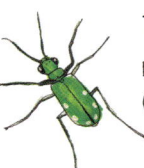

SIX-SPOTTED TIGER BEETLE

One out of every four animals on Earth is a beetle! Scientists believe there may be another 1 or 2 million species yet to be discovered.

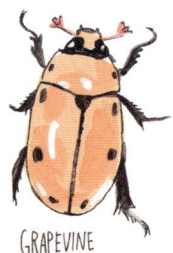

GRAPEVINE BEETLE

Living just about everywhere on the planet in a vast array of shapes and sizes, most adult beetles are easy to recognize by their armor, specifically their hardened forewings.

BURYING BEETLE

BLUE FUNGUS BEETLE

GOLDENROD SOLDIER BEETLE

PLEASING FUNGUS BEETLE

WASP BEETLE

STRIPED BLISTER BEETLE

GLORIOUS SCARAB BEETLE

HUHU BEETLE

0.5"   1"

The beetles on this page are their actual size.

FROG-LEGGED LEAF BEETLE

GOLDEN GROUND BEETLE

ANATOMY OF A BEETLE

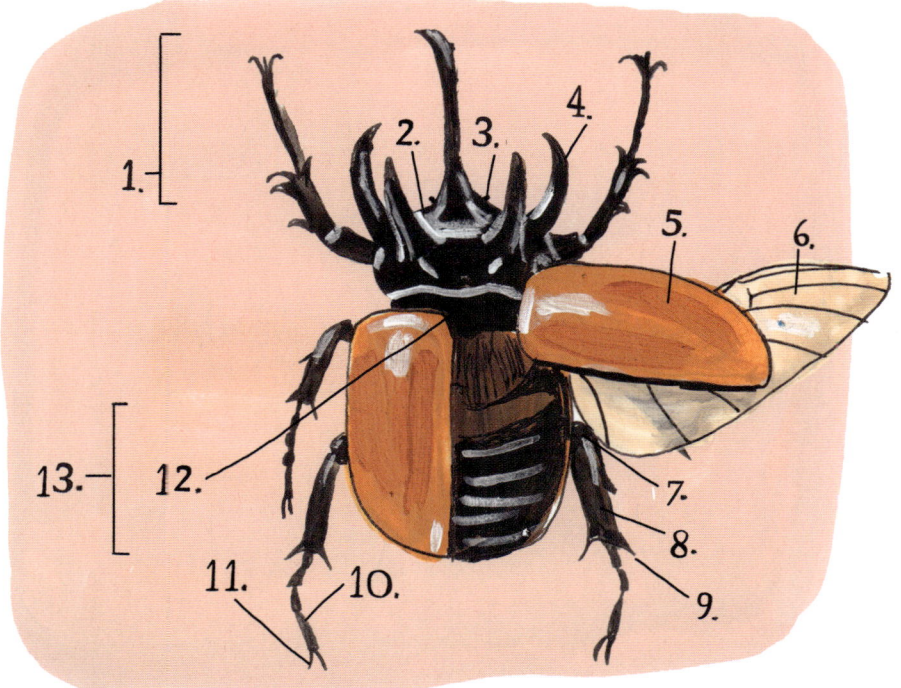

1. head
2. eye
3. antenna
4. horn
5. elytron
6. wing
7. femur
8. tibia
9. tibia spur
10. tarsus
11. tarsal claw
12. scutellum
13. abdomen

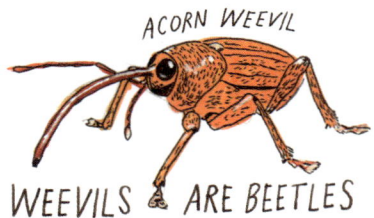

ACORN WEEVIL

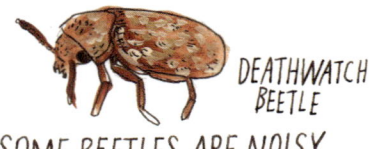

DEATHWATCH BEETLE

WEEVILS ARE BEETLES

They may look somewhat different with their long snouts, but they are still in the order Coleoptera. Like all beetles, weevils have mouthparts made for chewing, however theirs are found at the tip of the snout.

SOME BEETLES ARE NOISY

Like crickets and grasshoppers, many beetles also produce sounds by stridulation (rubbing parts of their bodies together). Some beetles will tap their abdomens on the ground to make sound. The deathwatch beetle bangs its head against wood! The sounds are sometimes for courtship, but also can be a form of defense to warn off an attacker.

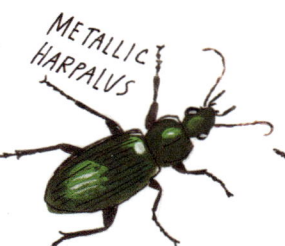

METALLIC HARPALUS

ROVE BEETLE

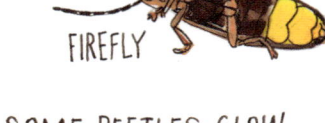

FIREFLY

EXPLOSIVE!

Ground beetles have glands in their abdomen that can eject an explosive and noxious gas capable of injuring small mammals such as shrews and killing other small invertebrates.

A BIG FAMILY

Rove beetles are the largest family in the beetle order, with more than 66,000 species. Fossils have been found dating back to the Triassic Era, 200 million years ago.

SOME BEETLES GLOW IN THE DARK

Fireflies (lightning bugs) and glow worms emit light by creating a reaction with a chemical in their bodies called luciferase. In the case of the firefly, the flashes are used to signal potential mates. For a glow worm, they are a means of attracting prey.

Lovely Ladies

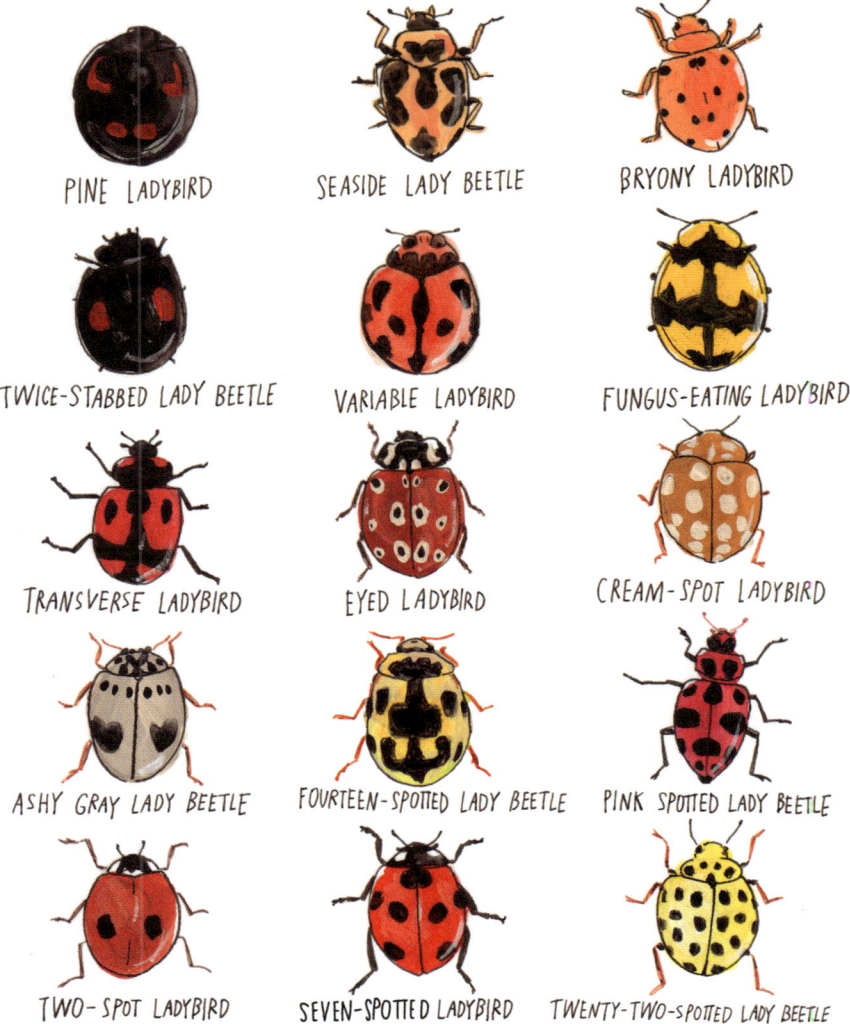

PINE LADYBIRD

SEASIDE LADY BEETLE

BRYONY LADYBIRD

TWICE-STABBED LADY BEETLE

VARIABLE LADYBIRD

FUNGUS-EATING LADYBIRD

TRANSVERSE LADYBIRD

EYED LADYBIRD

CREAM-SPOT LADYBIRD

ASHY GRAY LADY BEETLE

FOURTEEN-SPOTTED LADY BEETLE

PINK SPOTTED LADY BEETLE

TWO-SPOT LADYBIRD

SEVEN-SPOTTED LADYBIRD

TWENTY-TWO-SPOTTED LADY BEETLE

Entomologists refer to them as ladybug beetles (or ladybird beetles) to avoid confusion with true bugs. Farmers love these beetles because they eat aphids and other plant-eating insects. A single ladybug can eat up to 5,000 insects during its lifetime.

INSECTS ARE IMPORTANT

The world would not be the same without insects. Though often considered pests, insects provide food for birds, reptiles, and many other animals. Some pollinate flowers, and others help get rid of waste matter such as decaying feces or decaying flesh. They spin silk, which is turned into clothing, and they make honey, which sweetens our food.

Over the next 75 years, as many as half of the world's insect species may go extinct due to agriculture and human sprawl.

Insects such as bees, butterflies, and beetles are disappearing from ecosystems across the planet. Thankfully, people are making efforts to mitigate this disruption by avoiding pesticides and conserving their habitats.

POLLINATE THE PLANTS

More than 90 percent of the western bumble bee population disappeared in the past 20 years. Without bees and other insect pollinators we'd no longer have tomatoes, broccoli, melons, squash, or blueberries!

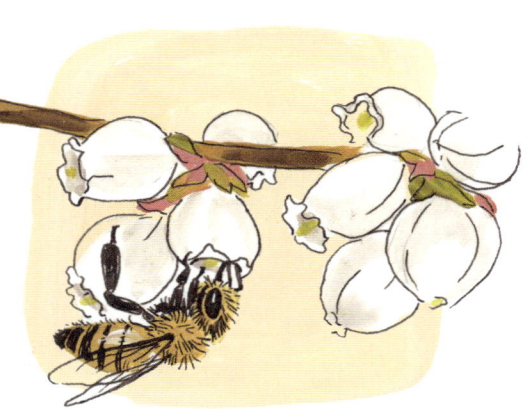

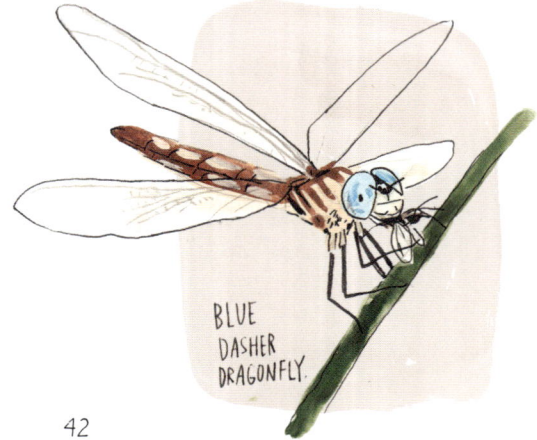

BLUE DASHER DRAGONFLY.

EAT THE PESTS

Mosquitoes can carry infectious diseases. Some dragonflies consume hundreds of mosquitoes in a single day.

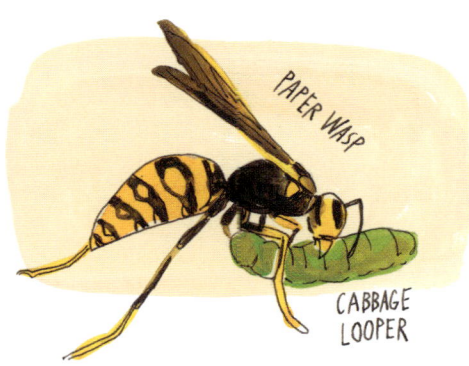

PAPER WASP

CABBAGE LOOPER

FIND THE BALANCE

While nobody wants to get stung by a wasp, we need them around to eat caterpillars, spiders, and crickets, keeping them from reaching pest levels.

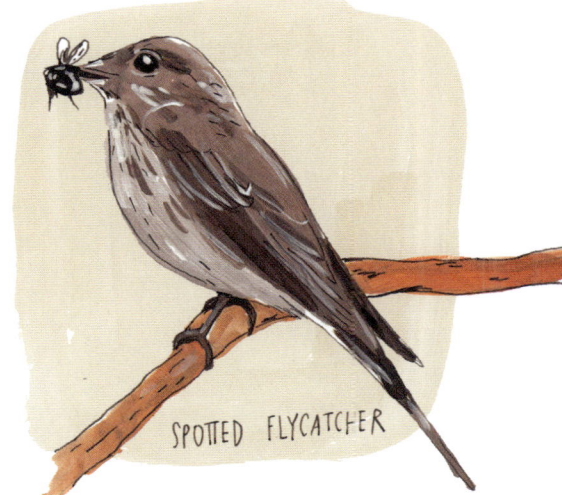

SPOTTED FLYCATCHER

FEED THE BIRDS

It has been estimated that birds consume approximately 500 tons of insects each year. Simply put: no insects, far fewer birds.

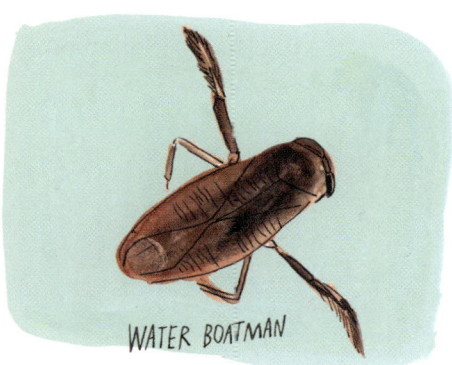

WATER BOATMAN

KEEP IT CLEAN

Insects contribute to the breakdown of debris in ponds, creeks, streams, and other wetlands. This helps keep water clean for plants, wildlife, and humans.

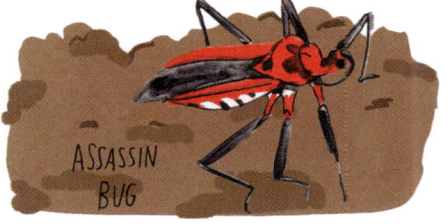

ASSASSIN BUG

CHURN THE SOIL

Only a very small percentage of insects damage crops. In fact, most are helpful by not only keeping the crop-eating insects in check, but also by breaking down and aerating the soil, which in turn helps plants grow.

CLIMATE CHANGE

As is the case with all living creatures, global warming and extreme weather events are having a profound effect on insects. Although incredibly tough, many insect species are being forced to relocate to cooler climates, which greatly impacts their life cycle, and even threatens extinction. This relocation also affects their interaction with other species, disrupting the food chain. Habitat destruction contributes to climate change and disrupts ecosystems.

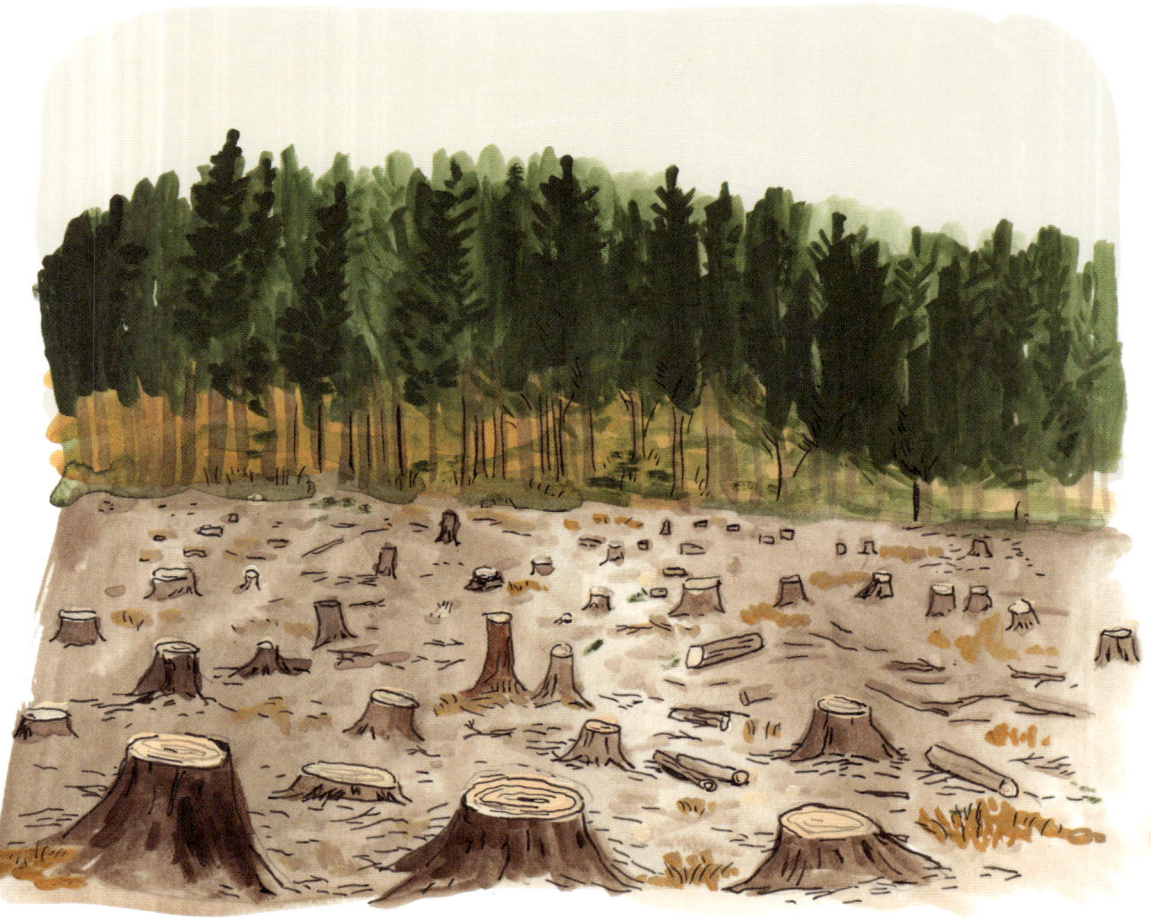

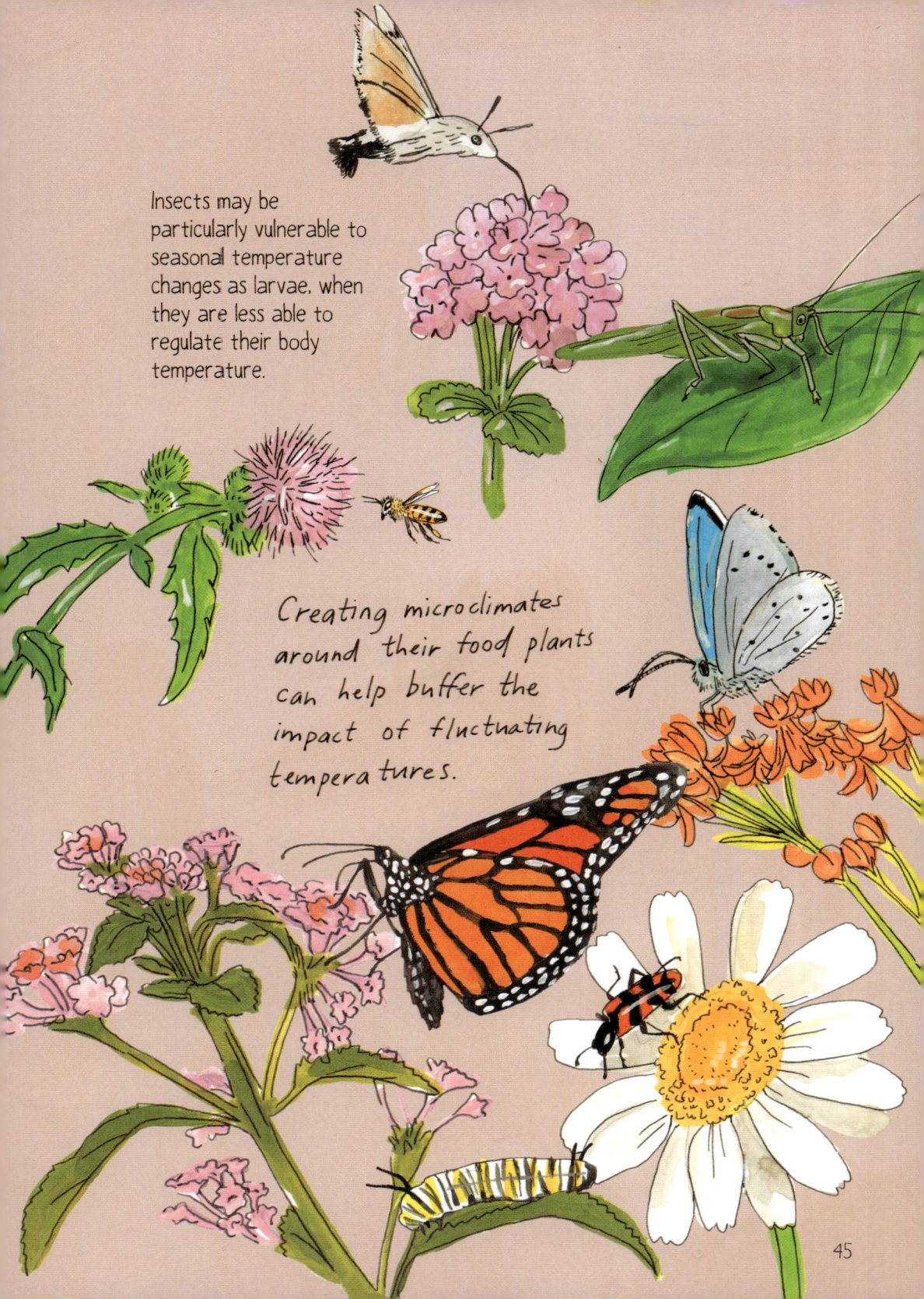

Insects may be particularly vulnerable to seasonal temperature changes as larvae, when they are less able to regulate their body temperature.

Creating microclimates around their food plants can help buffer the impact of fluctuating temperatures.

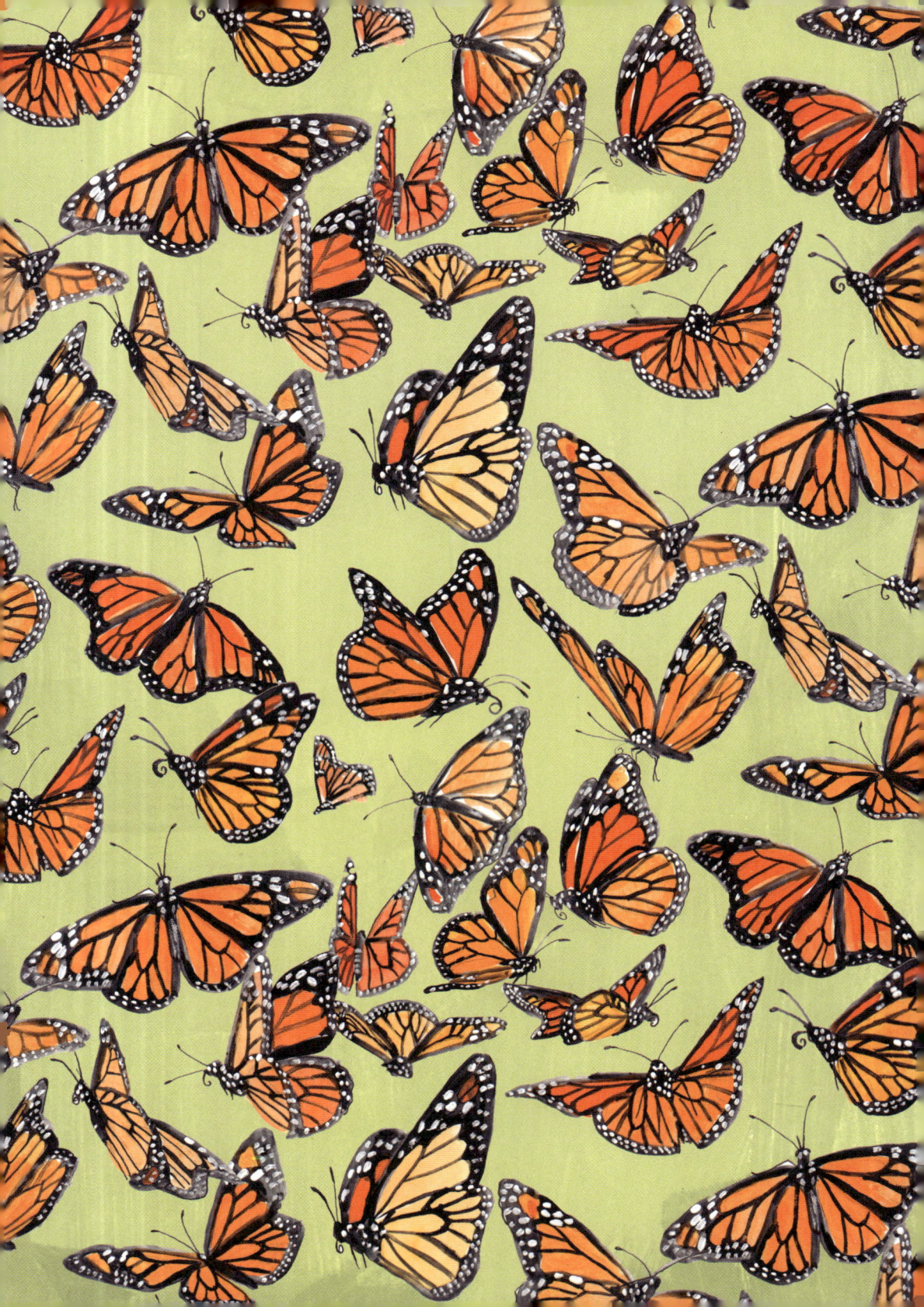

CHAPTER 2
Time Flies

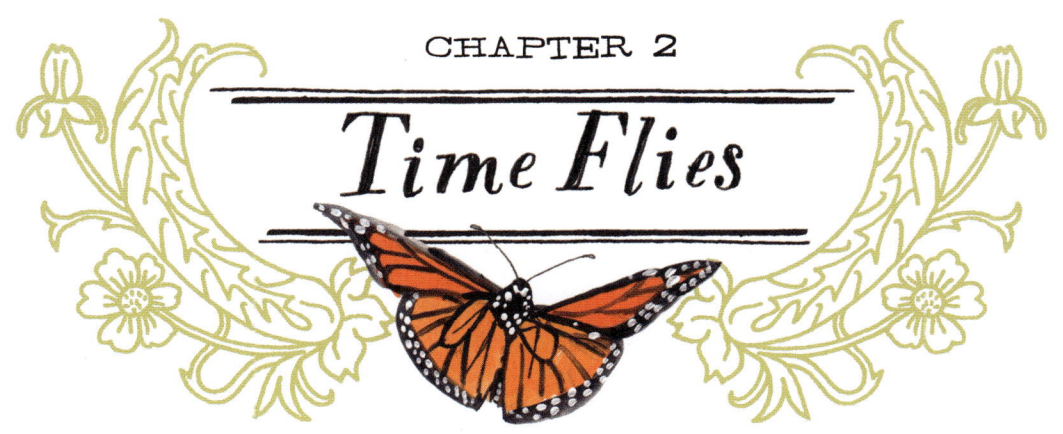

INSECT EGGS

Insects lay their eggs in a wide range of locations, such as on the undersides of leaves or in soil, water, wood, and sheltered spots like crevices and nests. Their eggs vary in shape, size, color, and texture for protection and to help them thrive in their specific habitats.

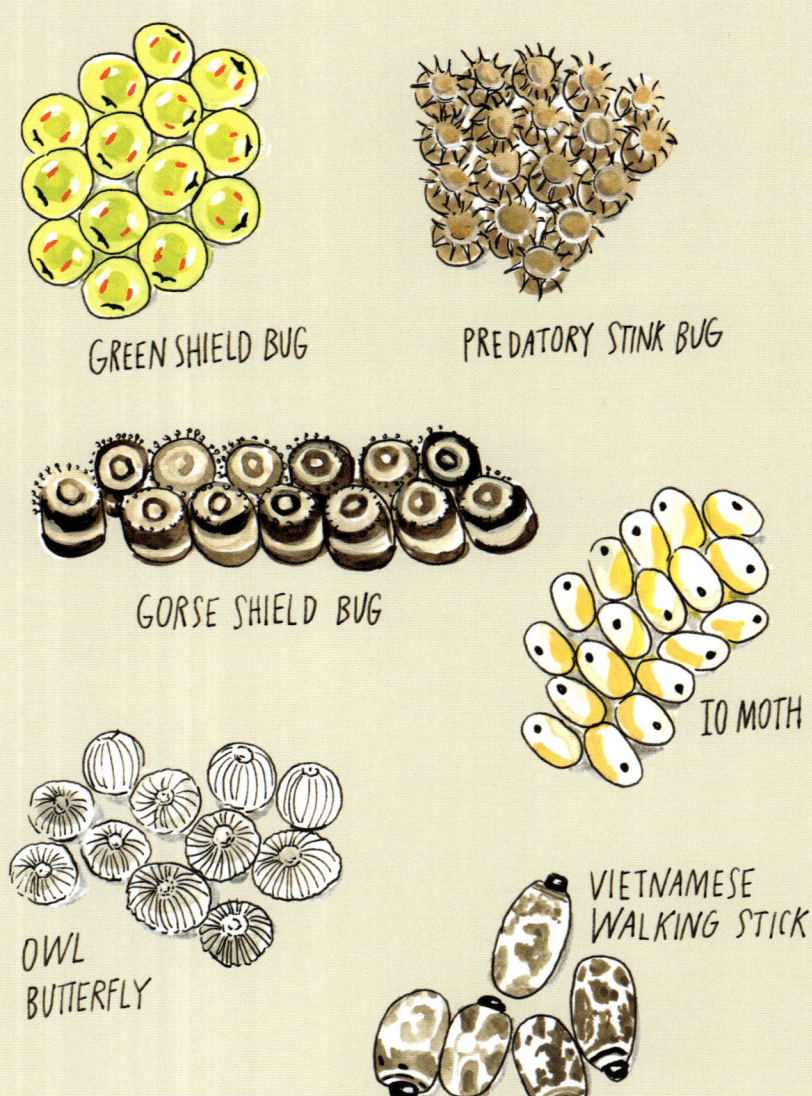

GREEN SHIELD BUG

PREDATORY STINK BUG

GORSE SHIELD BUG

IO MOTH

OWL BUTTERFLY

VIETNAMESE WALKING STICK

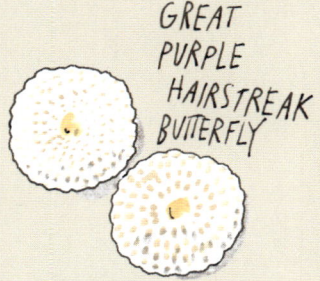

LADYBUG

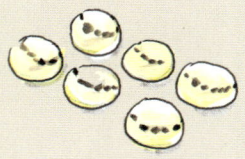

BLUE MORPHO BUTTERFLY

GREAT
PURPLE
HAIRSTREAK
BUTTERFLY

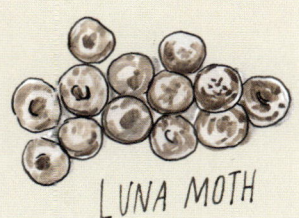

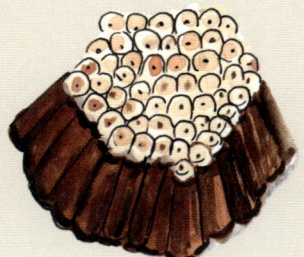

ASSASSIN BUG

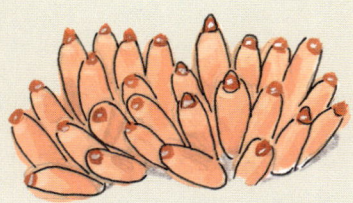

LUNA MOTH

ARGUS TORTOISE BEETLE

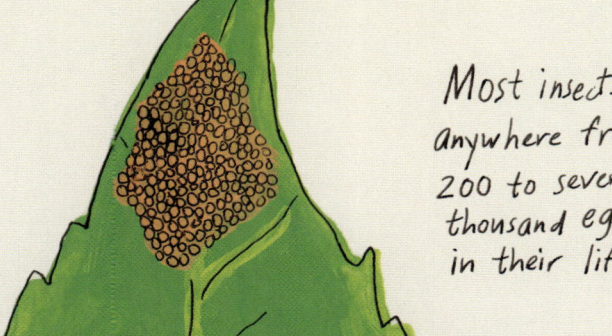

Most insects lay
anywhere from
200 to several
thousand eggs
in their lifetime.

INSECT GROWTH

Nearly all insects hatch from eggs and undergo metamorphosis, where the transformation from egg to adult takes place in distinct stages. Different species undergo different levels of metamorphosis.

NO METAMORPHOSIS

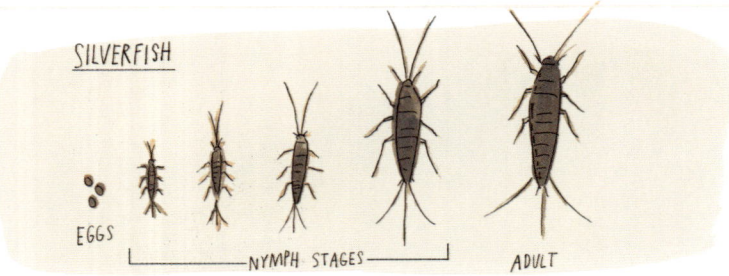

SILVERFISH

EGGS — NYMPH STAGES — ADULT

After hatching, the young look nearly the same as the adult, albeit smaller. As the insect grows and the exoskeleton becomes too tight, it occasionally molts, shedding its exoskeleton while a new layer takes its place. Simple metamorphosis is found in primitive wingless insects such as silverfish and springtails.

GRADUAL METAMORPHOSIS

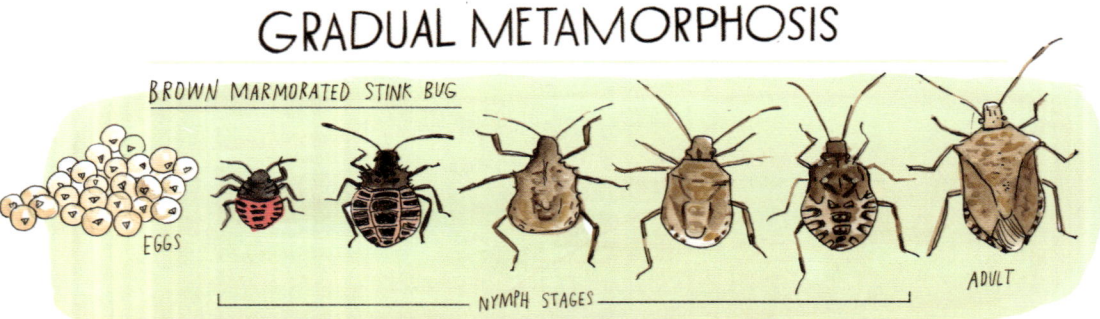

BROWN MARMORATED STINK BUG

EGGS — NYMPH STAGES — ADULT

This process, also called "incomplete metamorphosis," consists of three stages: egg, nymph, and adult. Some nymphs look similar to their adult counterparts, but others are completely different. Some, like dragonflies, have an underwater phase.

Nymphs are voracious eaters, quickly outgrowing their exoskeletons as they increase in size. The faster they grow, the sooner they can start the reproduction cycle all over again.

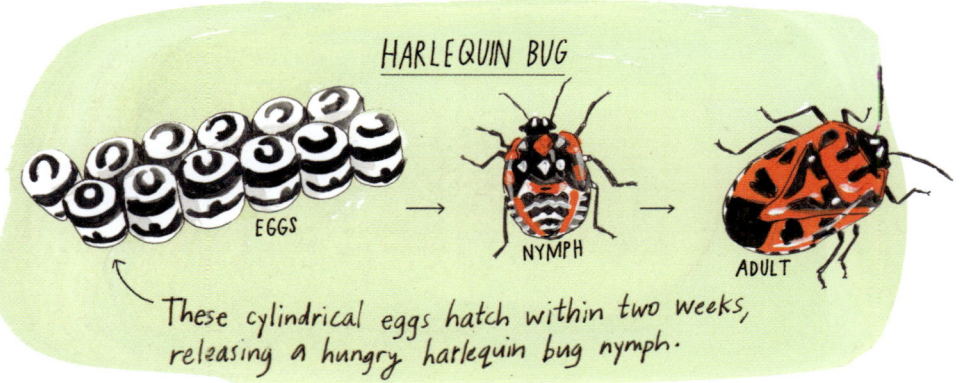

HARLEQUIN BUG

EGGS → NYMPH → ADULT

These cylindrical eggs hatch within two weeks, releasing a hungry harlequin bug nymph.

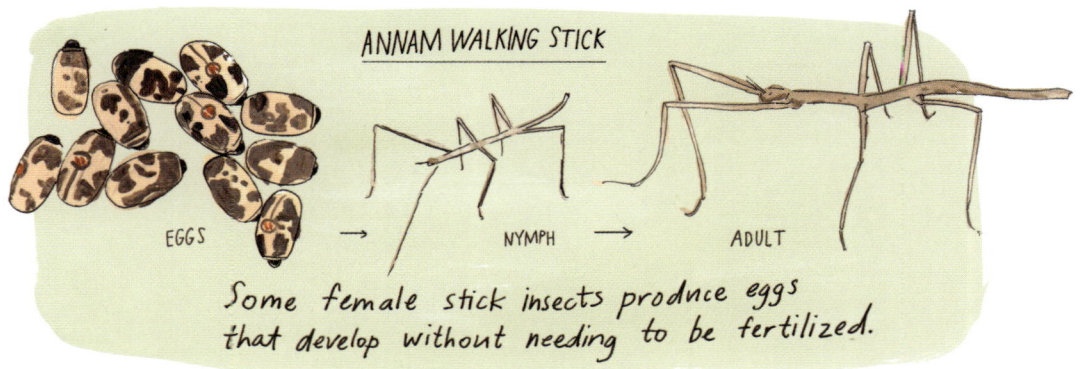

ANNAM WALKING STICK

EGGS → NYMPH → ADULT

Some female stick insects produce eggs that develop without needing to be fertilized.

MOLTING

Because of their hard-shell exoskeletons, molting (or shedding) is necessary for all insects to grow. As a new inner layer (cuticle) is formed, the outer layer splits and is pushed away as the insect wriggles free.

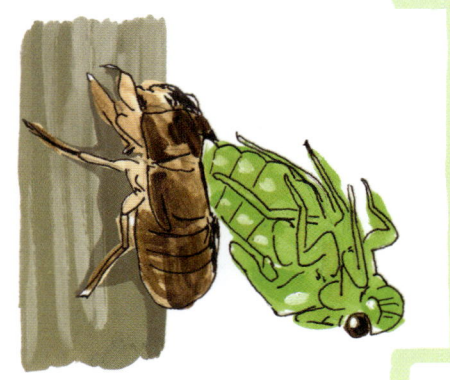

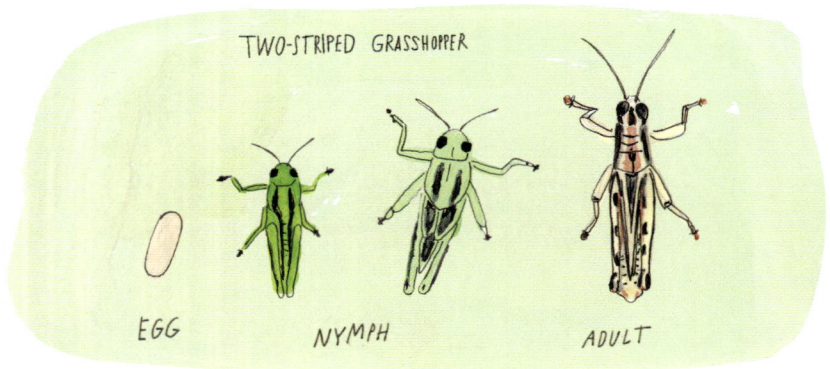

TWO-STRIPED GRASSHOPPER

EGG NYMPH ADULT

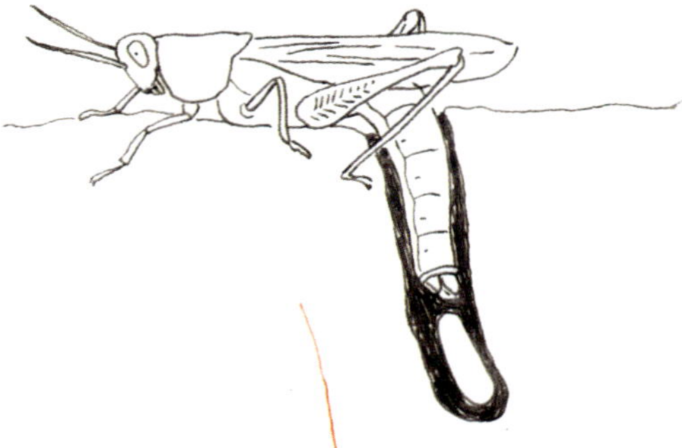

ONE WAY TO LAY EGGS

Using the tip of her abdomen, a female grasshopper digs a hole up to an inch deep. She lines the hole with a liquid, forming a pod, and deposits the eggs. Then she covers the pod with a frothy protective layer before burying it with dirt.

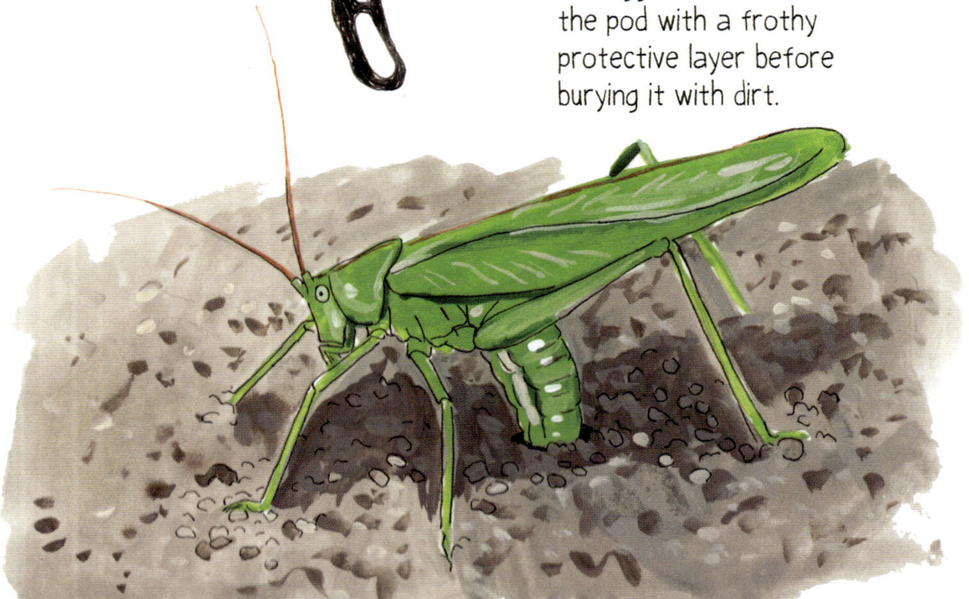

COMPLETE METAMORPHOSIS

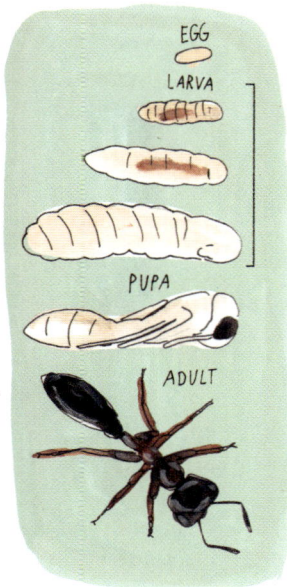

EGG

LARVA

PUPA

ADULT

The most complex development pattern is complete metamorphosis, where the insect goes through four distinct stages: egg, larva, pupa, and adult. Most insects, including butterflies, moths, ants, bees, wasps, and beetles, undergo this process.

Most insect larvae differ significantly from their adult form, typically having a worm-like appearance. They also generally lack reproductive organs and have feeding habits that are quite different from their full-grown counterparts.

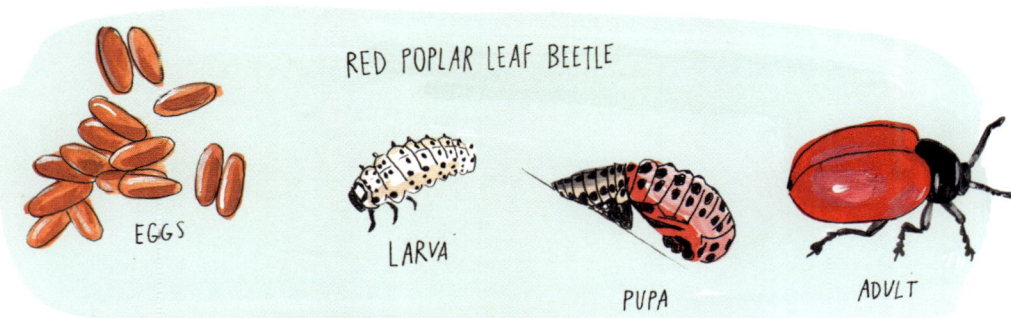

RED POPLAR LEAF BEETLE

EGGS

LARVA

PUPA

ADULT

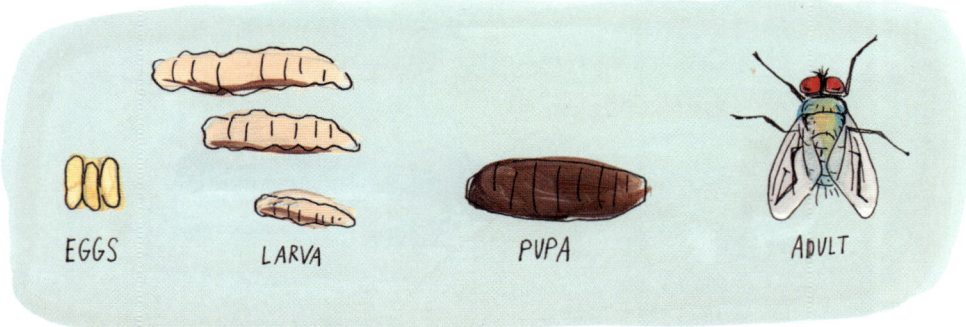

EGGS

LARVA

PUPA

ADULT

ANATOMY OF A DRAGONFLY

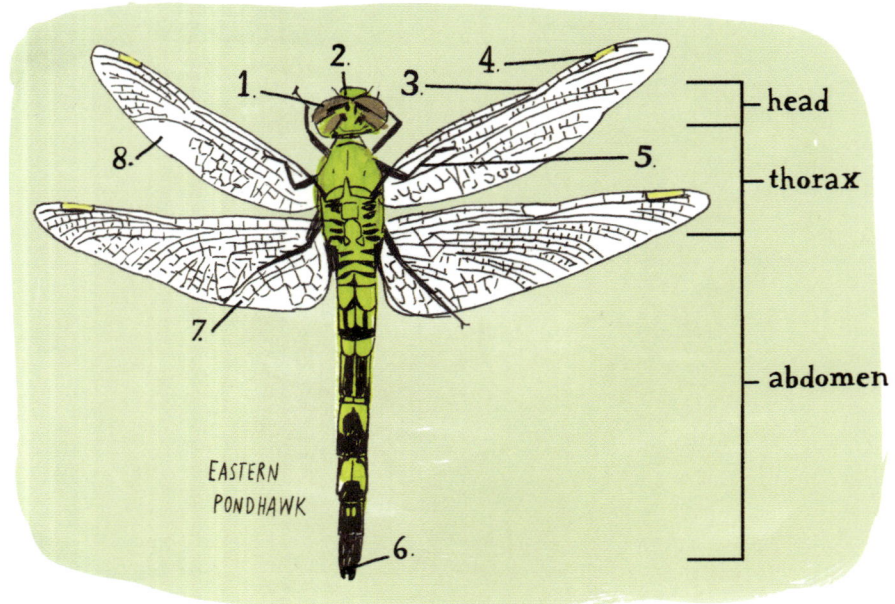

EASTERN PONDHAWK

head

thorax

abdomen

1. **compound eye** for hunting prey
2. **ocellus** used to detect light and orient during flight
3. **nodus** the microjoint located in the middle of the wing
4. **pterostigma** a heavier part of the wing that helps with gliding
5. **leg** covered in spiky hairs to help with gripping
6. **cerci** male dragonflies use these for holding females during mating
7. **hindwing**
8. **forewing**

Their wings beat 30 times per second and they are able to fly forward and backward, spin, and hover.

Life Cycle of a Dragonfly

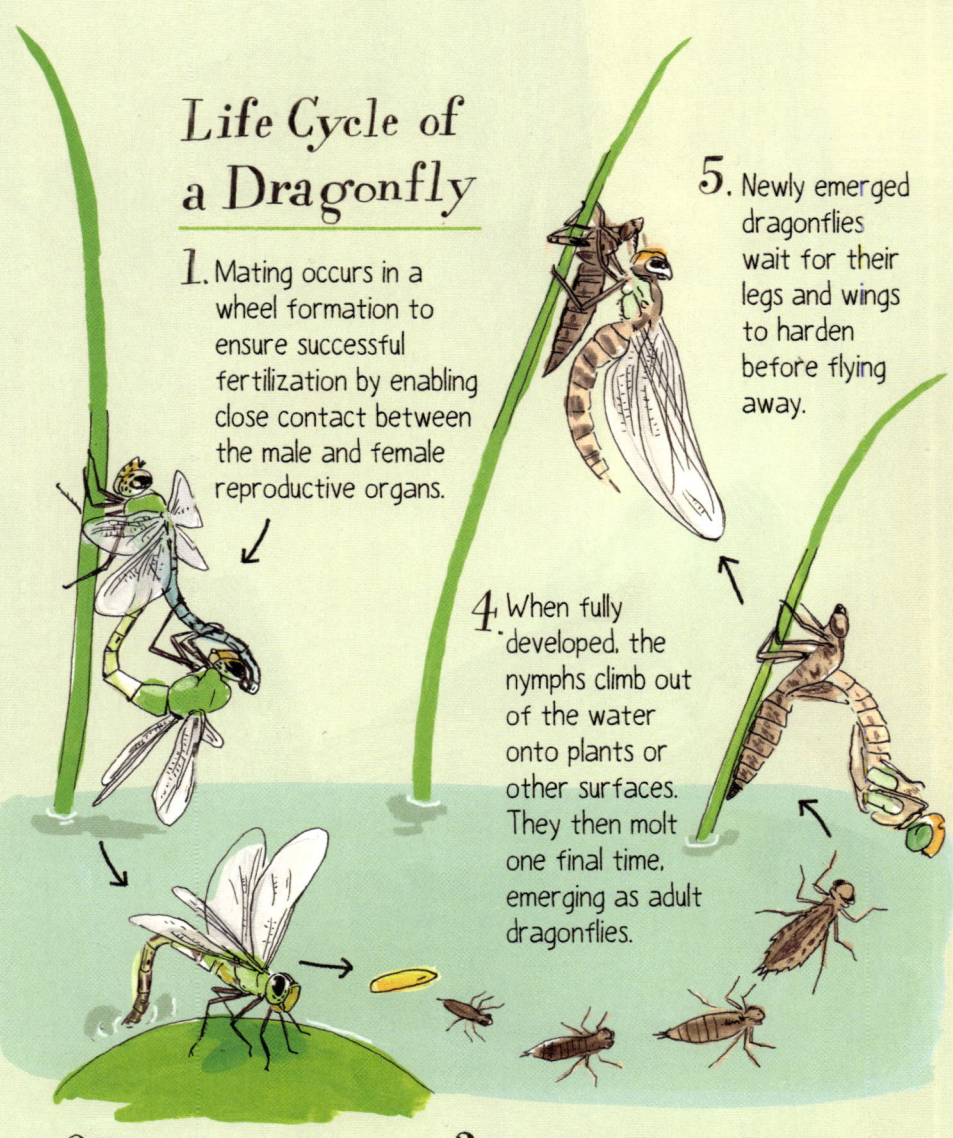

1. Mating occurs in a wheel formation to ensure successful fertilization by enabling close contact between the male and female reproductive organs.

2. Females deposit the fertilized eggs in water, often in clusters.

3. The eggs hatch into nymphs that live underwater for several months to several years, growing and molting multiple times. They are predatory and feed on other aquatic organisms.

4. When fully developed, the nymphs climb out of the water onto plants or other surfaces. They then molt one final time, emerging as adult dragonflies.

5. Newly emerged dragonflies wait for their legs and wings to harden before flying away.

AERIAL
ACROBATS

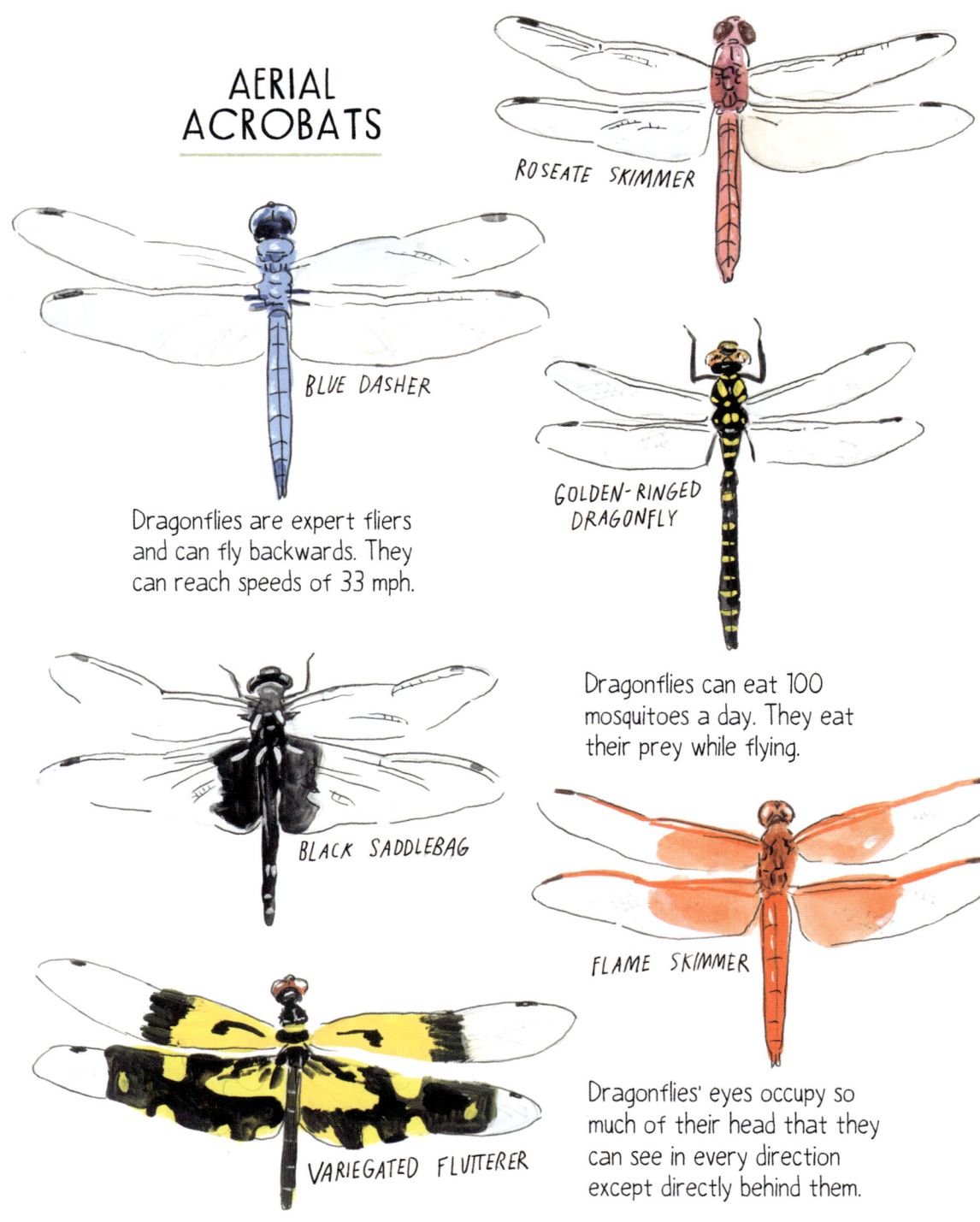

ROSEATE SKIMMER

BLUE DASHER

Dragonflies are expert fliers
and can fly backwards. They
can reach speeds of 33 mph.

GOLDEN-RINGED
DRAGONFLY

BLACK SADDLEBAG

Dragonflies can eat 100
mosquitoes a day. They eat
their prey while flying.

FLAME SKIMMER

VARIEGATED FLUTTERER

Dragonflies' eyes occupy so
much of their head that they
can see in every direction
except directly behind them.

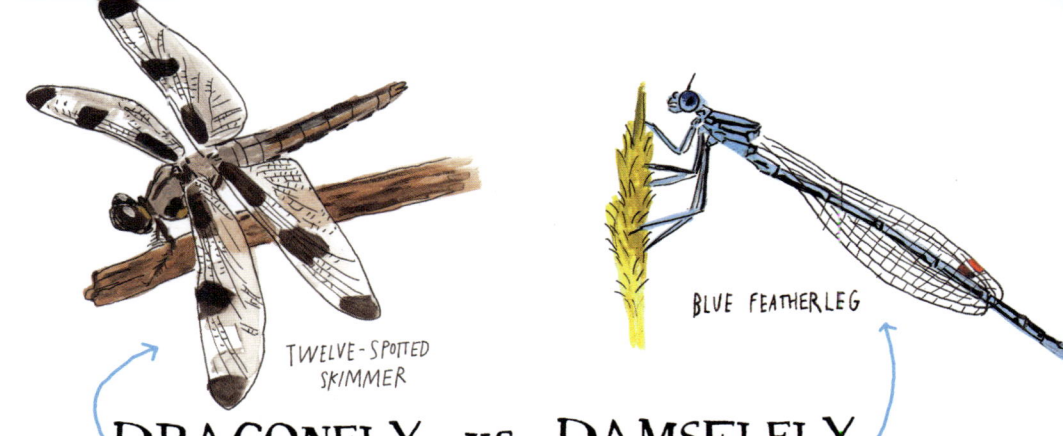

TWELVE-SPOTTED SKIMMER

BLUE FEATHERLEG

DRAGONFLY vs. DAMSELFLY

BULKY, THICK BODY	LONG, SLENDER BODY
WINGS BROAD AT BASE	WINGS NARROW AT BASE
WINGS OPEN AT REST	WINGS CLOSED AT REST
LARGE EYES THAT TOUCH OR ARE VERY CLOSE	LARGE EYES THAT DON'T TOUCH – GAP BETWEEN
STRONG, FAST FLIERS	WEAKER, MORE FLUTTERY FLIERS
FLY OVER LARGE BODIES OF WATER	FOUND CLOSER TO EDGES OF WATER BODIES

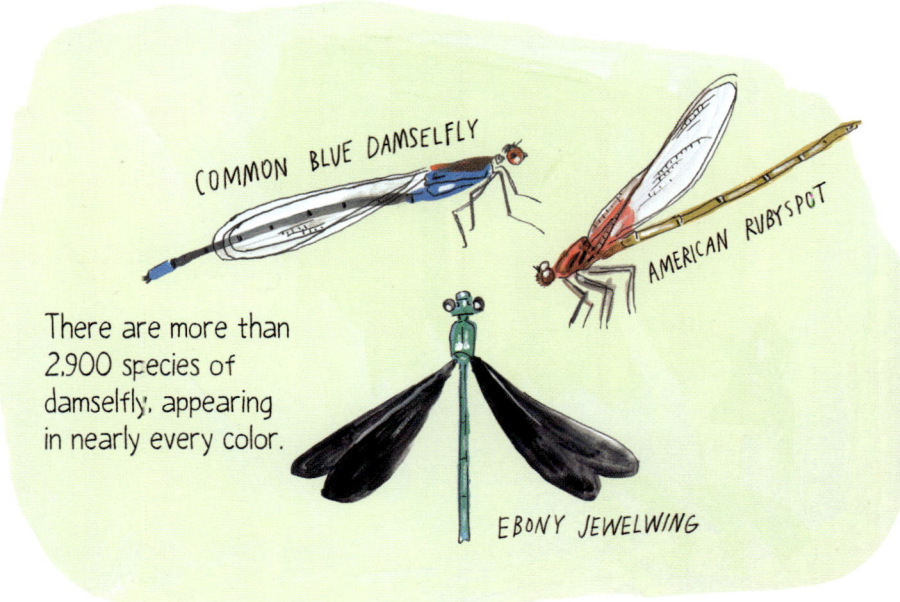

COMMON BLUE DAMSELFLY

AMERICAN RUBYSPOT

There are more than 2,900 species of damselfly, appearing in nearly every color.

EBONY JEWELWING

ANATOMY OF A CICADA

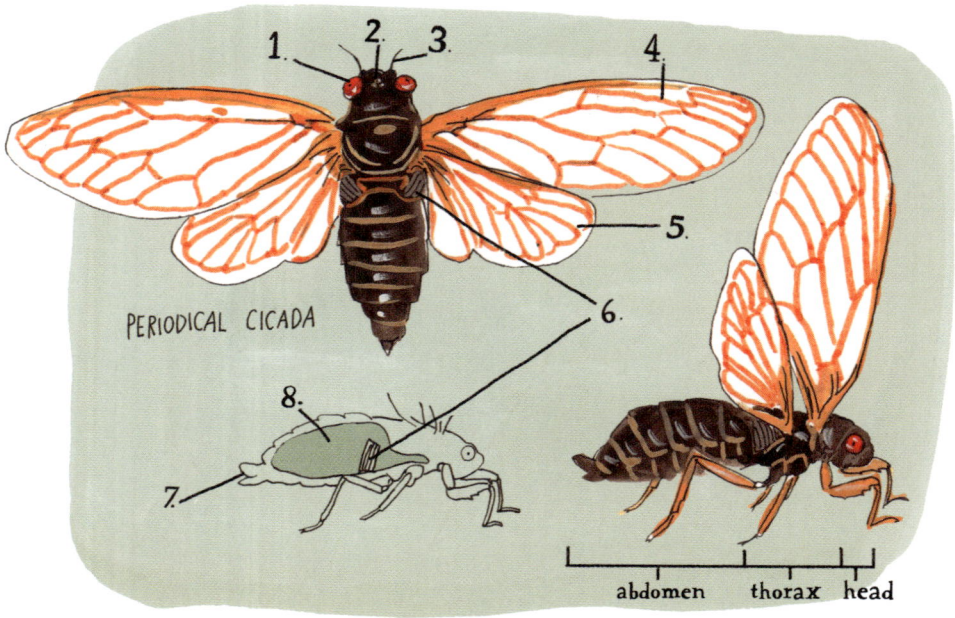

PERIODICAL CICADA

abdomen   thorax   head

1. compound eye
2. ocellus
3. antenna
4. forewing
5. hindwing
6. tymbal (male)
7. ovipositor (female)
8. air sac

The tymbals on the abdomen of a male cicada are ribbed membranes that engage powerful muscles to create their loud buzzing. Cicadas' bodies have big air sacs that act as echoing sound chambers.

THE LIFE CYCLE OF A PERIODICAL CICADA

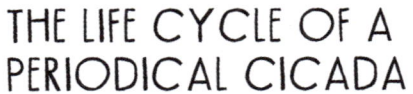

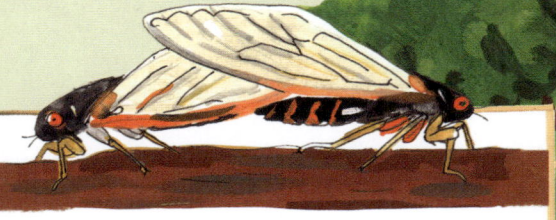

1. Using their tymbals, adult males "sing" to attract females for mating.

2. The female cuts crevices in tree branches and inserts about 20 eggs into each of them using her ovipositor.

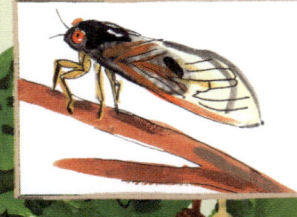

7. The adults are ready to find their mates.

3. The eggs hatch in 6-8 weeks. Tiny nymphs fall to the ground and burrow underneath the soil.

6. It molts one last time, emerging pale but eventually darkening and hardening.

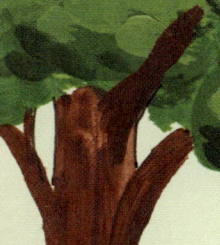

5. The nymph digs its way up out of the ground.

4. Underground, nymphs feed off of tree roots. They stay here 13-17 years, molting many times.

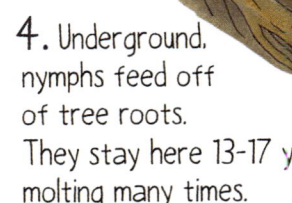

The Four Stages of a Moth or Butterfly

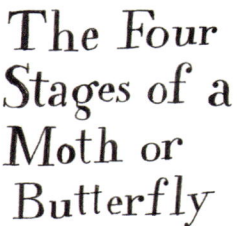

CECROPIA MOTH

1.

Eggs are usually laid on plants that the newborn caterpillars will eat.

↓

2.

Caterpillars eat a lot and grow a lot, shedding their skin several times as they increase in size.

→

3. Inside the cocoon (moth) or chrysalis (butterfly), the caterpillar metamorphosizes into an adult. This can take a week or as long as a year depending on the species.

↑

4. The adults mate, the female lays eggs, and the cycle starts all over again.

← MONARCH BUTTERFLY CATERPILLAR

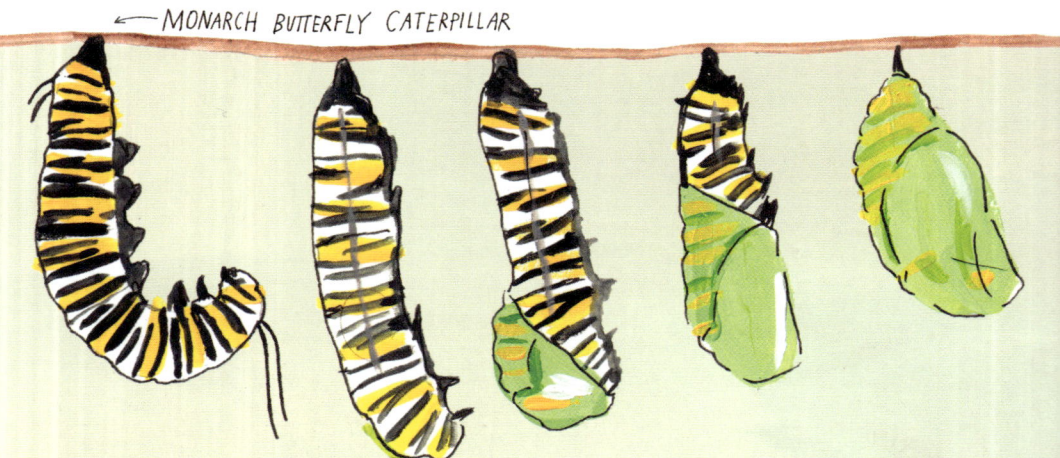

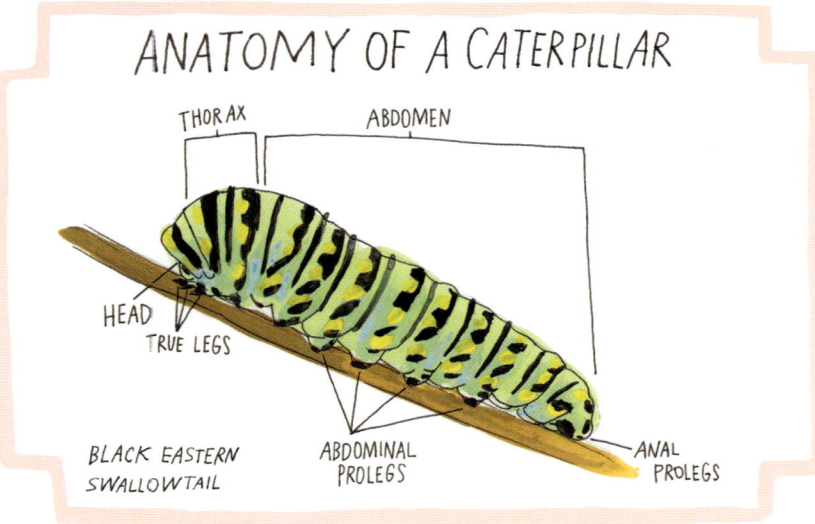

ANATOMY OF A CATERPILLAR

THORAX

ABDOMEN

HEAD

TRUE LEGS

BLACK EASTERN
SWALLOWTAIL

ABDOMINAL
PROLEGS

ANAL
PROLEGS

There are more than 180,000 species of caterpillar, which are the larval stage of butterflies and moths. Caterpillars have a single mission: to eat and grow. Some caterpillars grow as much as 1,000 times before metamorphizing into a butterfly or moth. Most caterpillars eat plants, but some eat bugs and even other caterpillars.

Once inside the chrysalis, the caterpillar's body turns to a soupy substance before reforming into a butterfly.

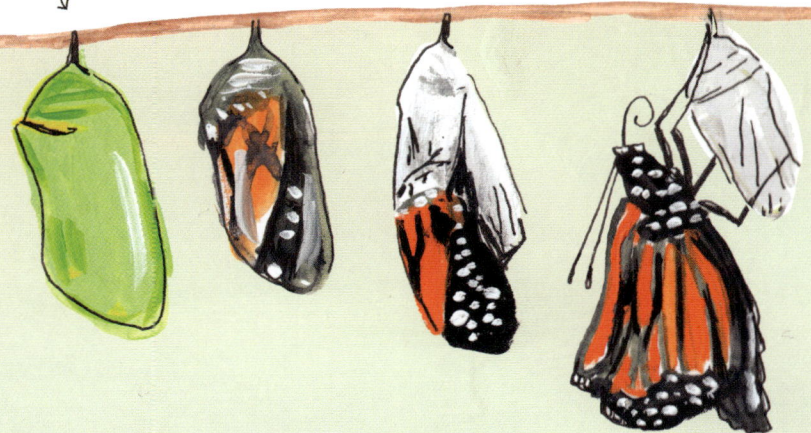

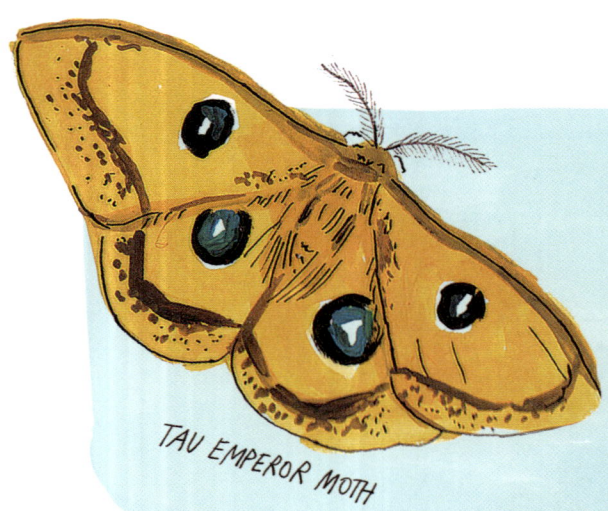

MOTHS

- ARE NOCTURNAL
- HAVE FEATHERY ANTENNAE
- KEEP THEMSELVES WARM BY FLAPPING THEIR WINGS
- CAN LIVE UP TO TEN MONTHS (SOME SPECIES)
- MAKE A COCOON

TAU EMPEROR MOTH

IS IT A MOTH OR A BUTTERFLY?

Both moths and butterflies belong to the order Lepidoptera. There are nearly 180,000 known species of Lepidoptera, around 90 percent of which are moths. Butterflies and moths have three body parts, four wings, and a pair of antennae; however, there are several key differences between them.

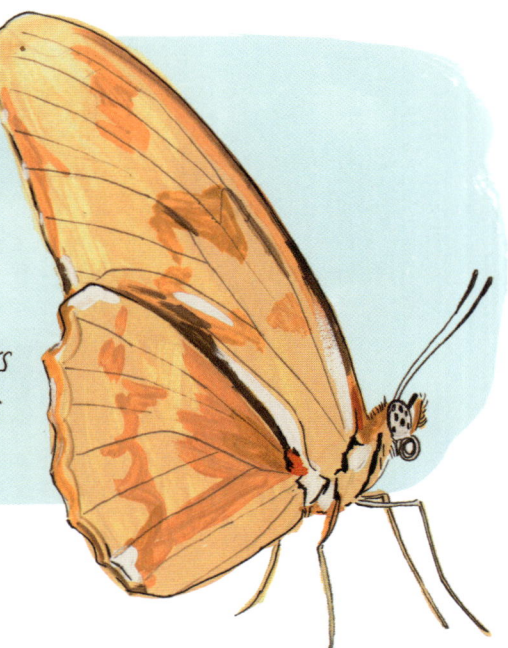

BUTTERFLIES

- ARE PRIMARILY ACTIVE DURING THE DAY
- HAVE SLENDER, CLUB-SHAPED ANTENNAE
- MOST LIVE LESS THAN TWO WEEKS
- MAKE A HANGING CHRYSALIS

ANATOMY OF A BUTTERFLY

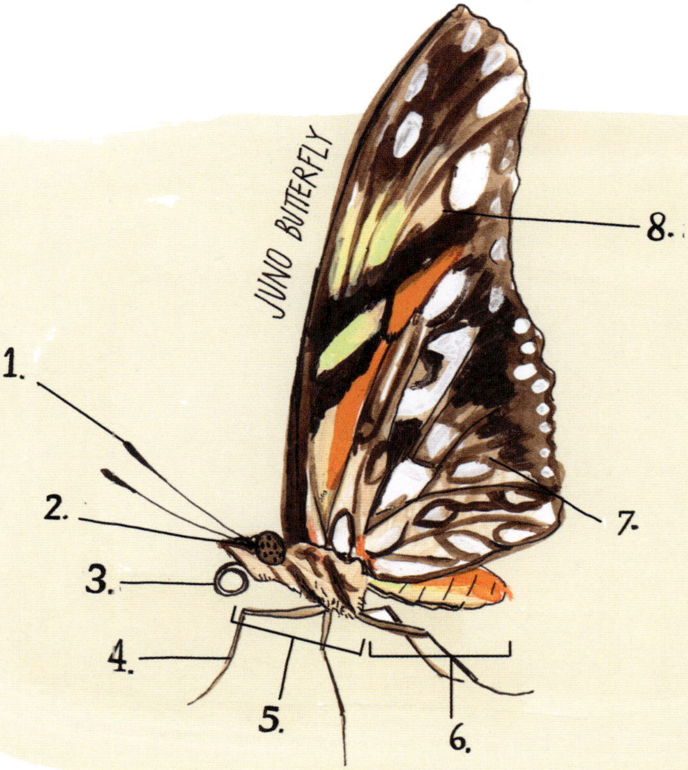

JUNO BUTTERFLY

8.

1.

2.

3.

4.

5.

6.

7.

1. **antennae** used to taste and smell and sense moisture and temperature
2. **compound eyes** each one has up to 1,700 ommatidia (light receptors and lenses)
3. **proboscis** like a long straw for feeding and drinking
4. **legs** three pairs (Nymphalids hold up the front pair and stand on the other four.)
5. **thorax** three body segments that contain the flight muscles
6. **abdomen** contains the digestive system, respiratory equipment, heart, and sex organs
7. **forewing** ⎤
8. **hindwing** ⎦ — two pairs of overlapping wings that flap and sometimes glide

BUTTERFLY CHRYSALISES

Chrysalises come in all different colors and shapes to camouflage with their surroundings and avoid predators—some mimic leaves, while others use metallic colors to reflect light and confuse attackers.

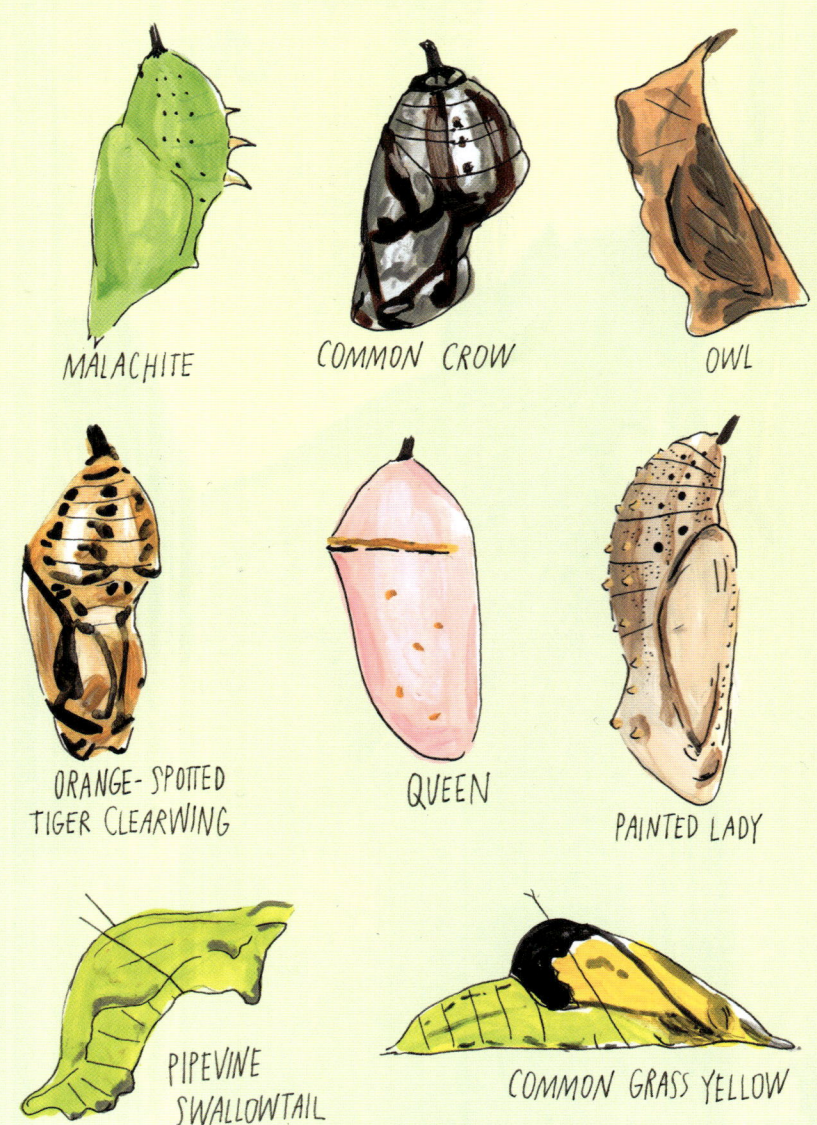

MALACHITE

COMMON CROW

OWL

ORANGE-SPOTTED TIGER CLEARWING

QUEEN

PAINTED LADY

PIPEVINE SWALLOWTAIL

COMMON GRASS YELLOW

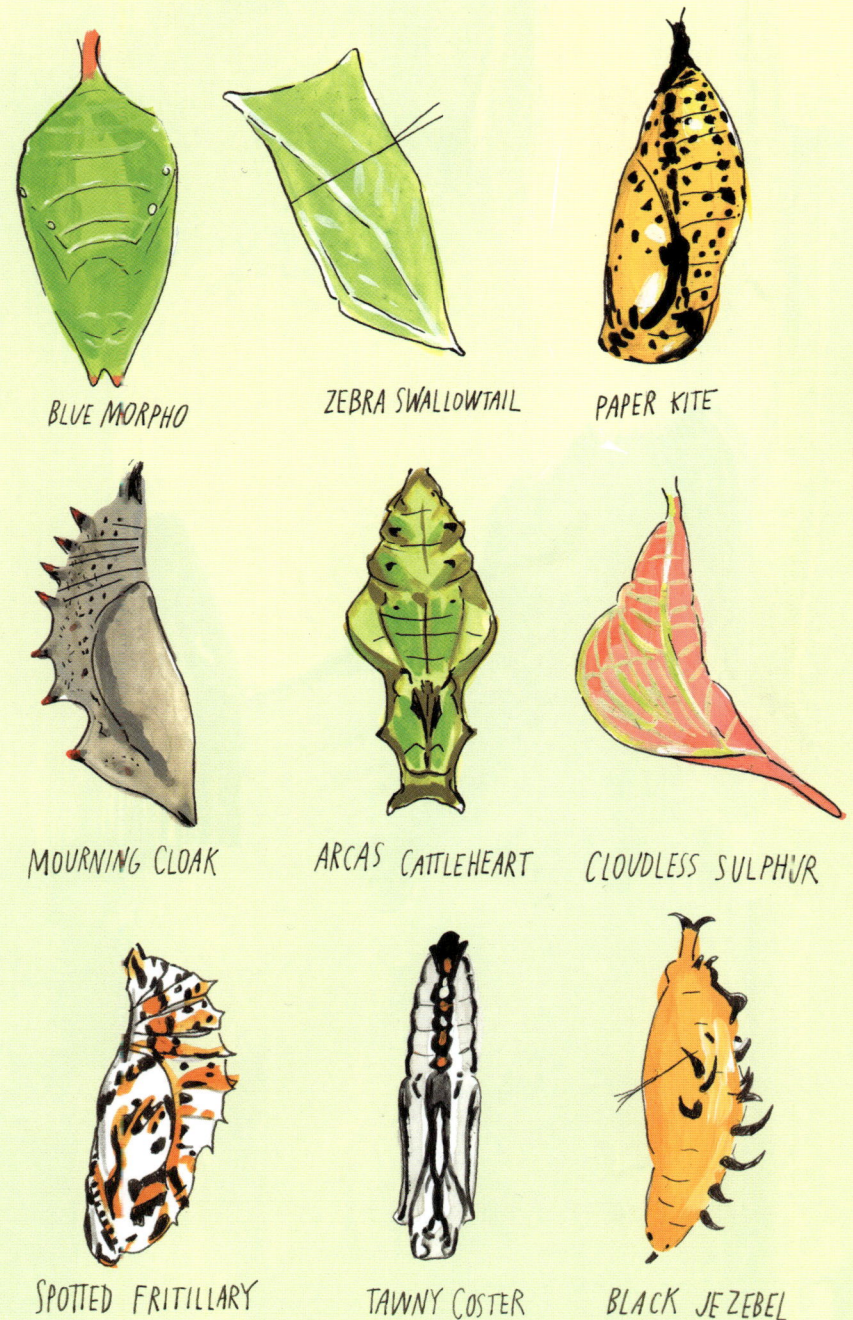

BLUE MORPHO

ZEBRA SWALLOWTAIL

PAPER KITE

MOURNING CLOAK

ARCAS CATTLEHEART

CLOUDLESS SULPHUR

SPOTTED FRITILLARY

TAWNY COSTER

BLACK JEZEBEL

THIS GROWS INTO THAT

Caterpillars are as varied as the adult moths and butterflies they become. They also generally lack reproductive organs and have feeding habits that are quite different from their full-grown counterparts.

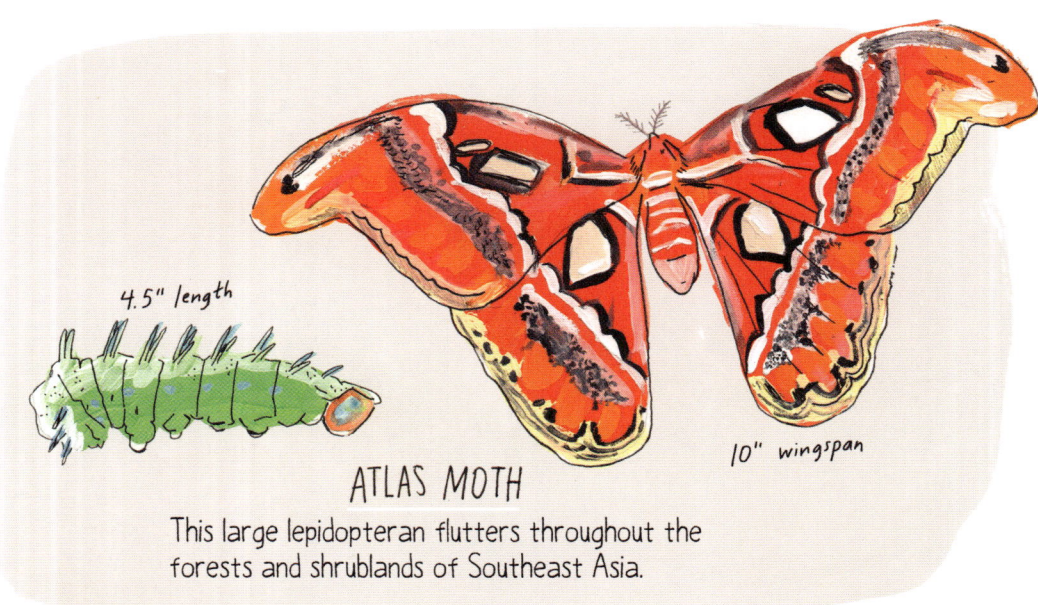

4.5" length

10" wingspan

ATLAS MOTH

This large lepidopteran flutters throughout the forests and shrublands of Southeast Asia.

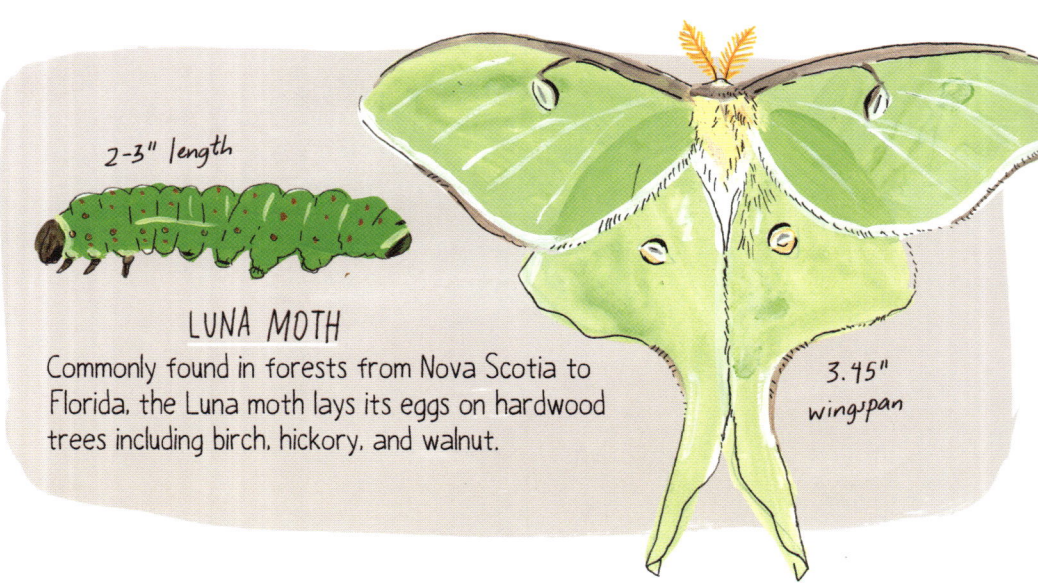

2-3" length

LUNA MOTH

Commonly found in forests from Nova Scotia to Florida, the Luna moth lays its eggs on hardwood trees including birch, hickory, and walnut.

3.45" wingspan

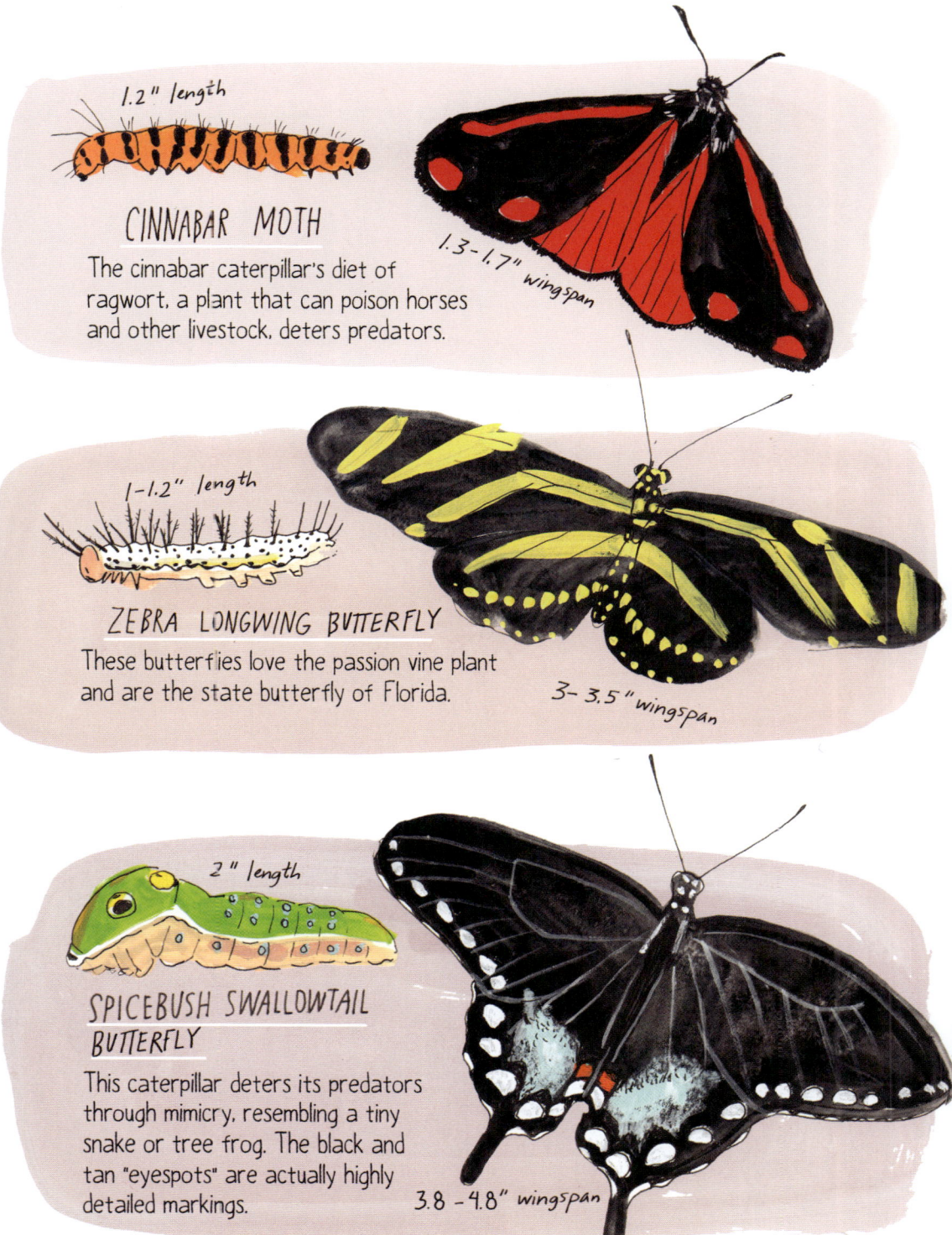

1.2" length

CINNABAR MOTH

The cinnabar caterpillar's diet of ragwort, a plant that can poison horses and other livestock, deters predators.

1.3-1.7" wingspan

1-1.2" length

ZEBRA LONGWING BUTTERFLY

These butterflies love the passion vine plant and are the state butterfly of Florida.

3-3.5" wingspan

2" length

SPICEBUSH SWALLOWTAIL BUTTERFLY

This caterpillar deters its predators through mimicry, resembling a tiny snake or tree frog. The black and tan "eyespots" are actually highly detailed markings.

3.8-4.8" wingspan

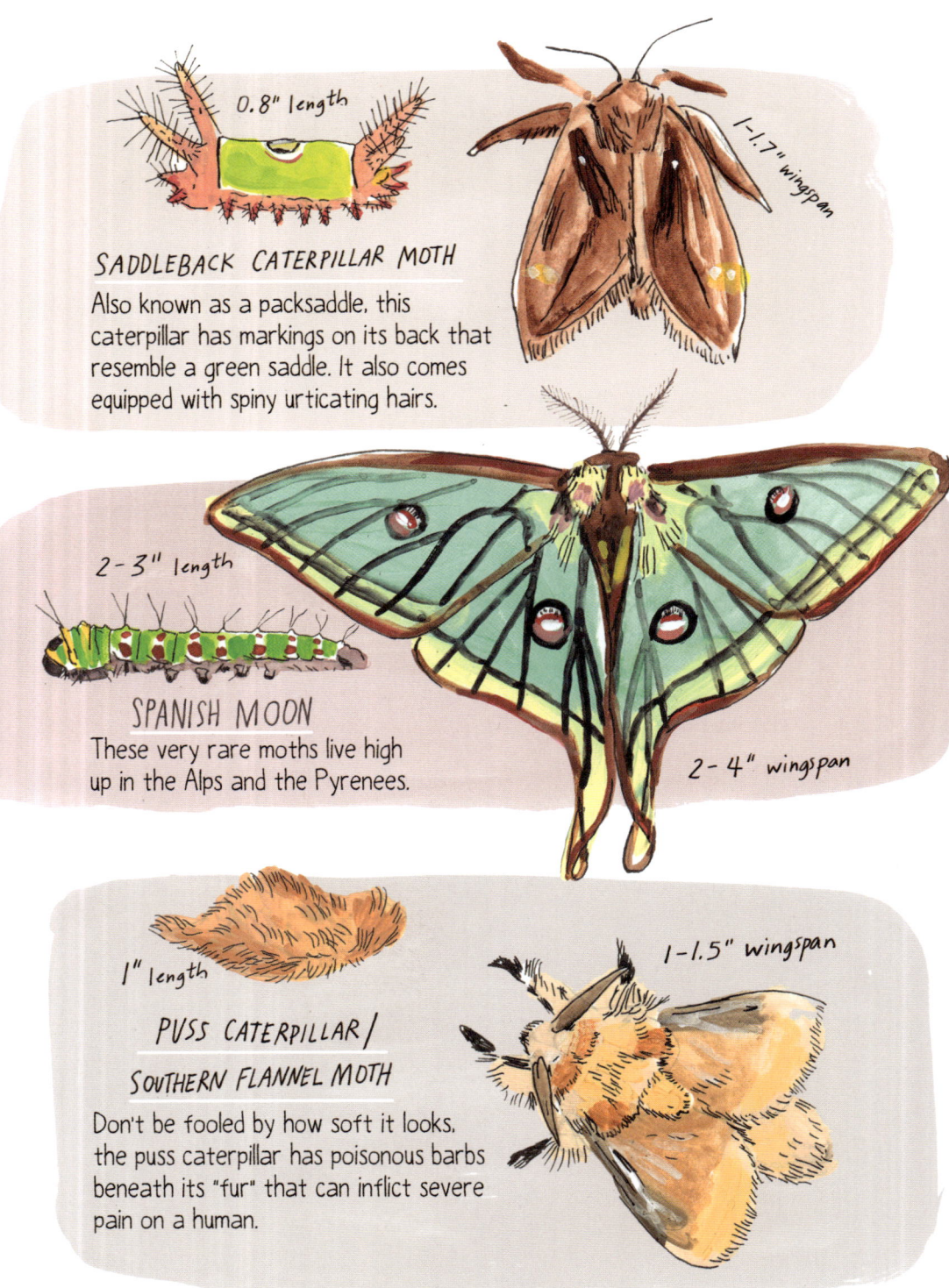

0.8" length

1-1.7" wingspan

SADDLEBACK CATERPILLAR MOTH

Also known as a packsaddle, this caterpillar has markings on its back that resemble a green saddle. It also comes equipped with spiny urticating hairs.

2-3" length

SPANISH MOON

These very rare moths live high up in the Alps and the Pyrenees.

2-4" wingspan

1" length

1-1.5" wingspan

PUSS CATERPILLAR /
SOUTHERN FLANNEL MOTH

Don't be fooled by how soft it looks, the puss caterpillar has poisonous barbs beneath its "fur" that can inflict severe pain on a human.

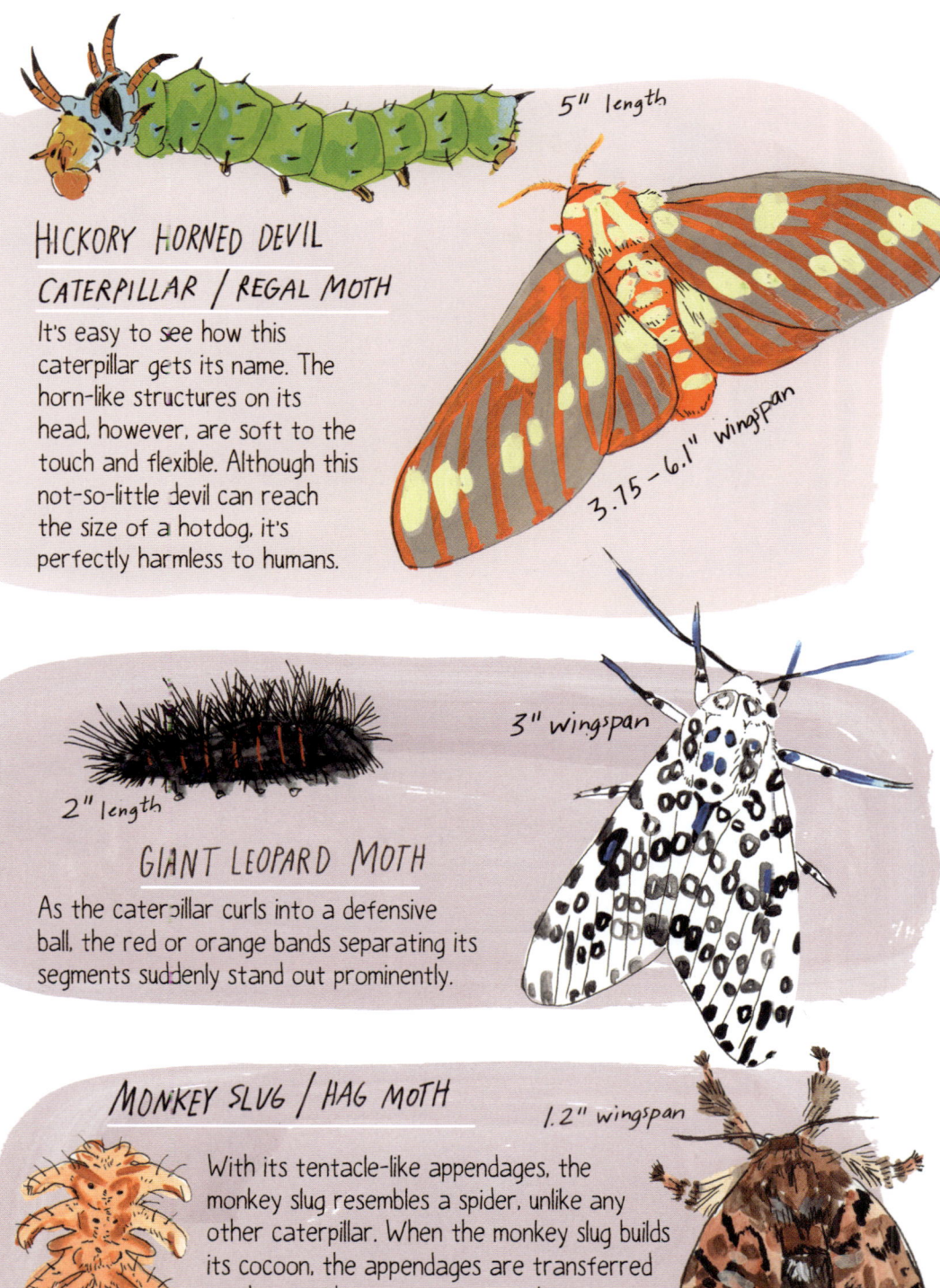

5" length

HICKORY HORNED DEVIL
CATERPILLAR / REGAL MOTH

It's easy to see how this caterpillar gets its name. The horn-like structures on its head, however, are soft to the touch and flexible. Although this not-so-little devil can reach the size of a hotdog, it's perfectly harmless to humans.

3.75 – 6.1" wingspan

2" length

3" wingspan

GIANT LEOPARD MOTH

As the caterpillar curls into a defensive ball, the red or orange bands separating its segments suddenly stand out prominently.

MONKEY SLUG / HAG MOTH

1.2" wingspan

With its tentacle-like appendages, the monkey slug resembles a spider, unlike any other caterpillar. When the monkey slug builds its cocoon, the appendages are transferred to the outside to act as camouflage.

0.6 – 0.9" length

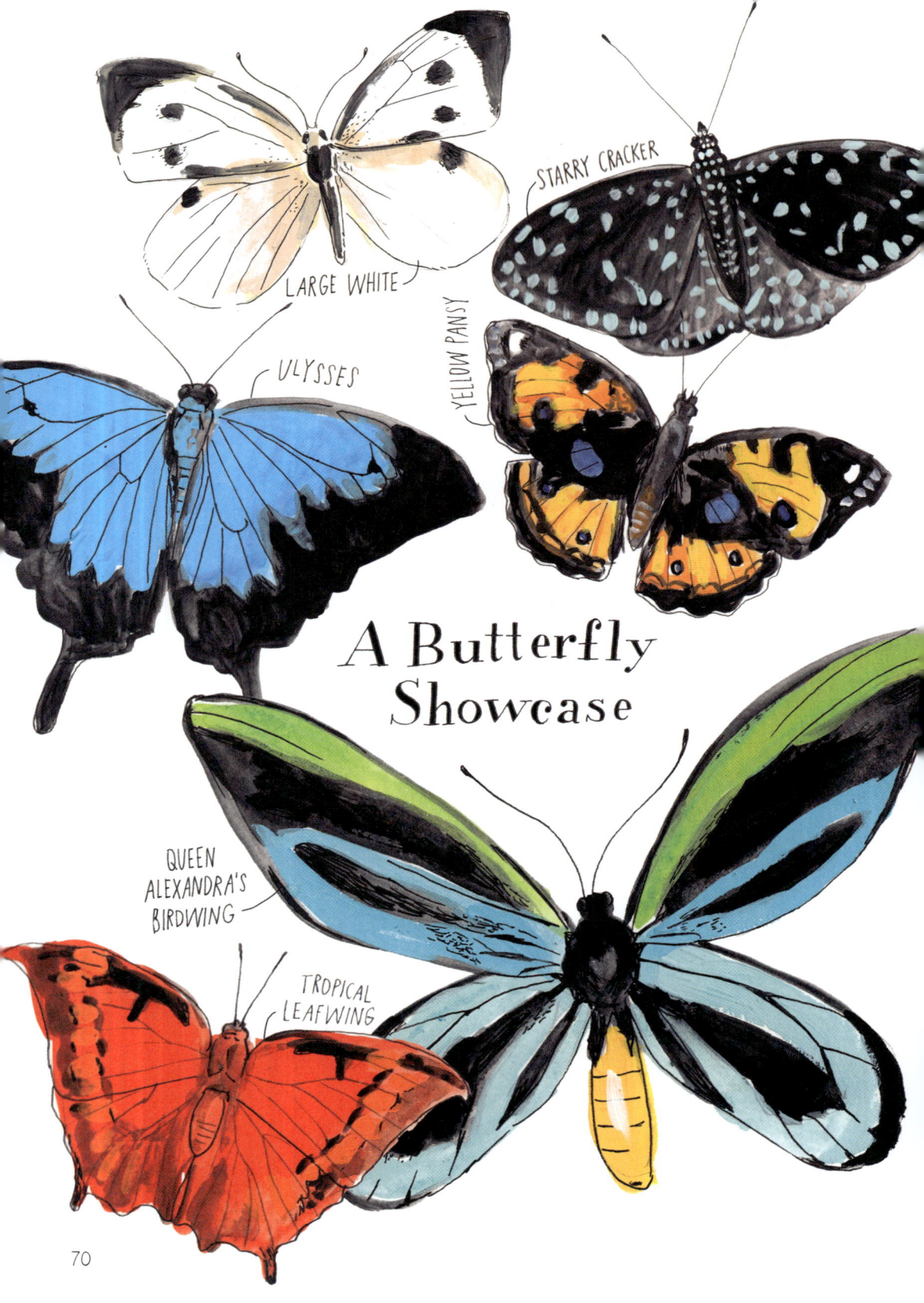

STARRY CRACKER

LARGE WHITE

ULYSSES

YELLOW PANSY

A Butterfly Showcase

QUEEN ALEXANDRA'S BIRDWING

TROPICAL LEAFWING

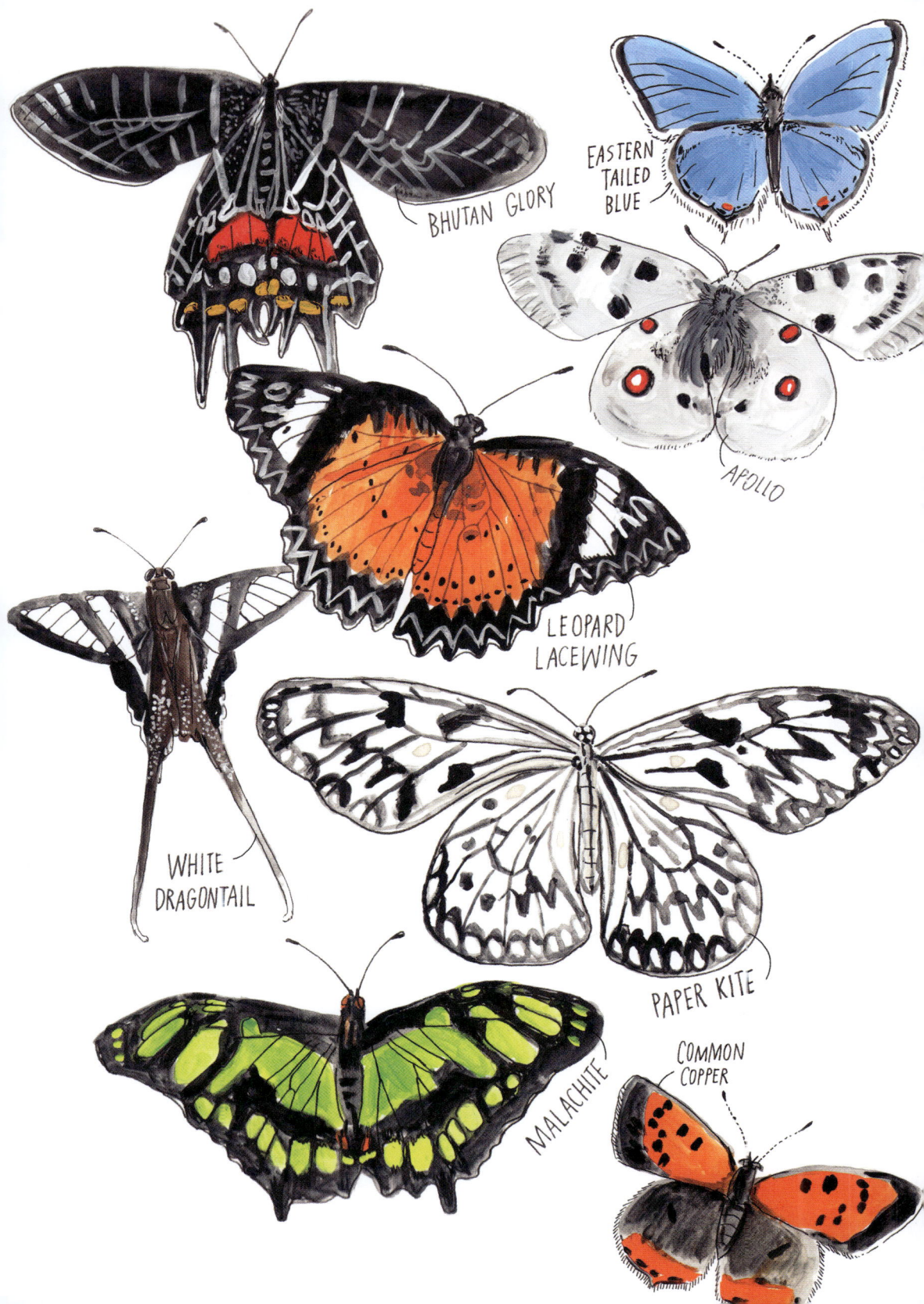

BHUTAN GLORY

EASTERN
TAILED
BLUE

APOLLO

LEOPARD
LACEWING

WHITE
DRAGONTAIL

PAPER KITE

MALACHITE

COMMON
COPPER

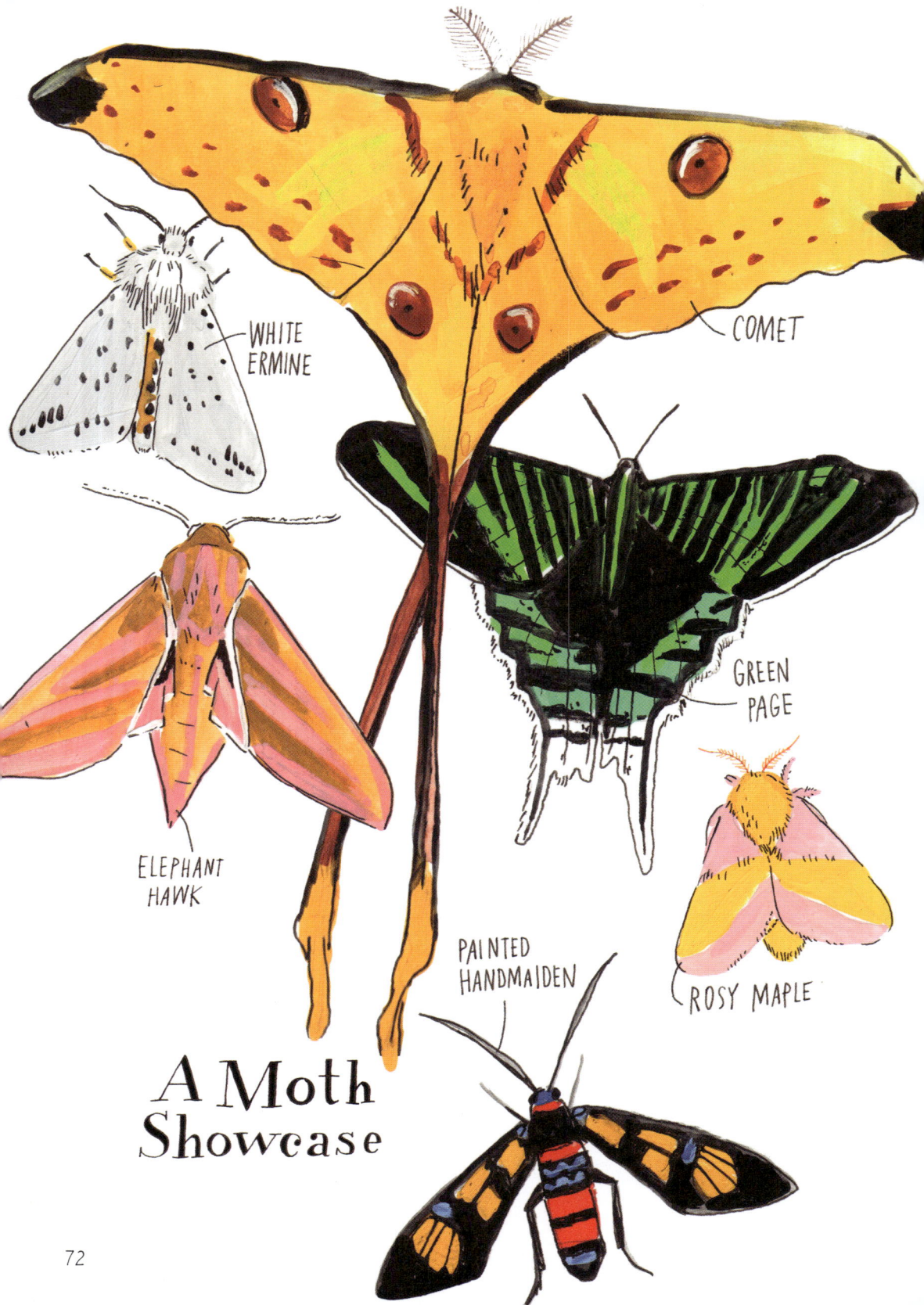

WHITE ERMINE

COMET

GREEN PAGE

ELEPHANT HAWK

PAINTED HANDMAIDEN

ROSY MAPLE

A Moth Showcase

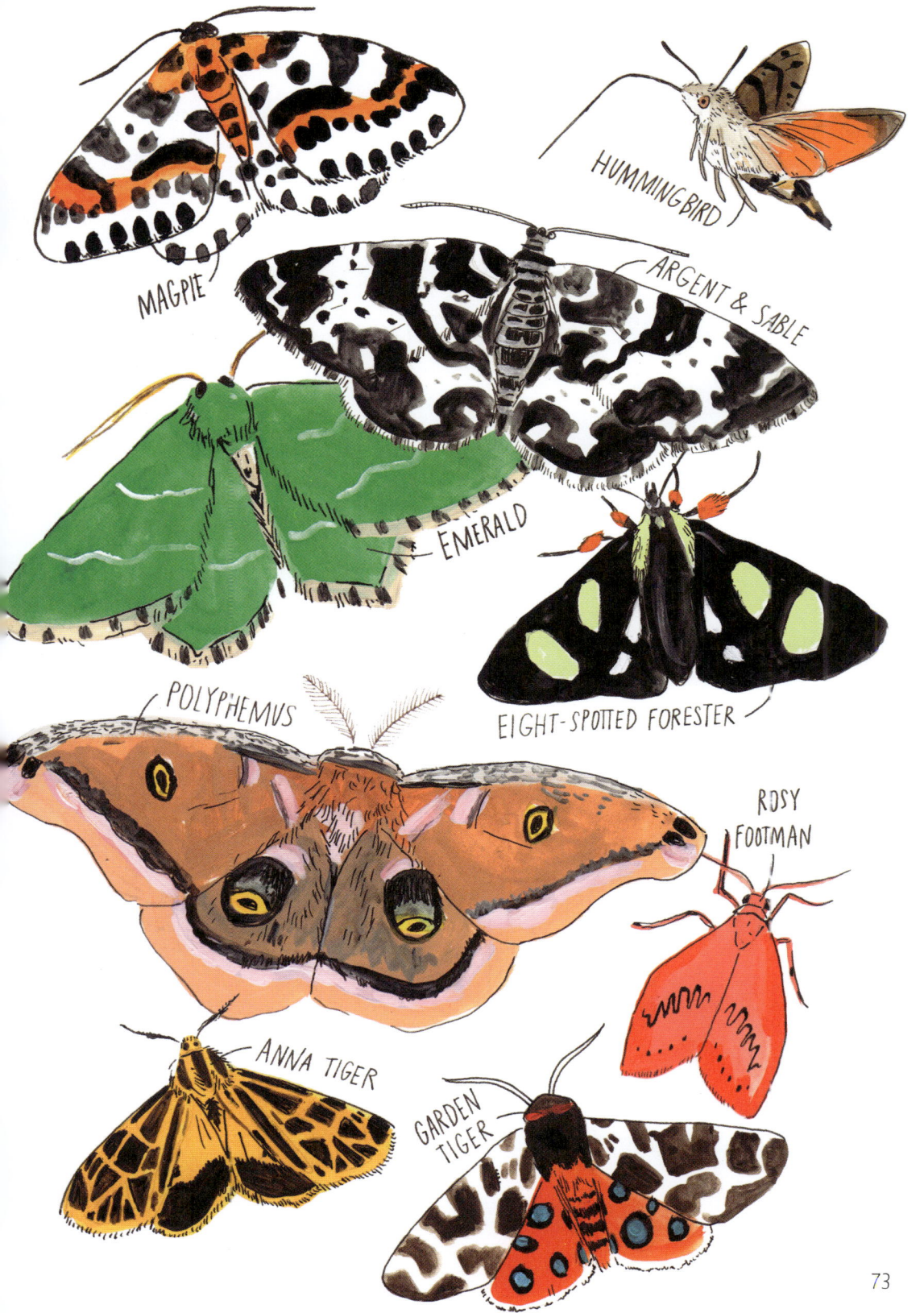

MAGPIE

HUMMINGBIRD

ARGENT & SABLE

EMERALD

EIGHT-SPOTTED FORESTER

POLYPHEMUS

ROSY FOOTMAN

ANNA TIGER

GARDEN TIGER

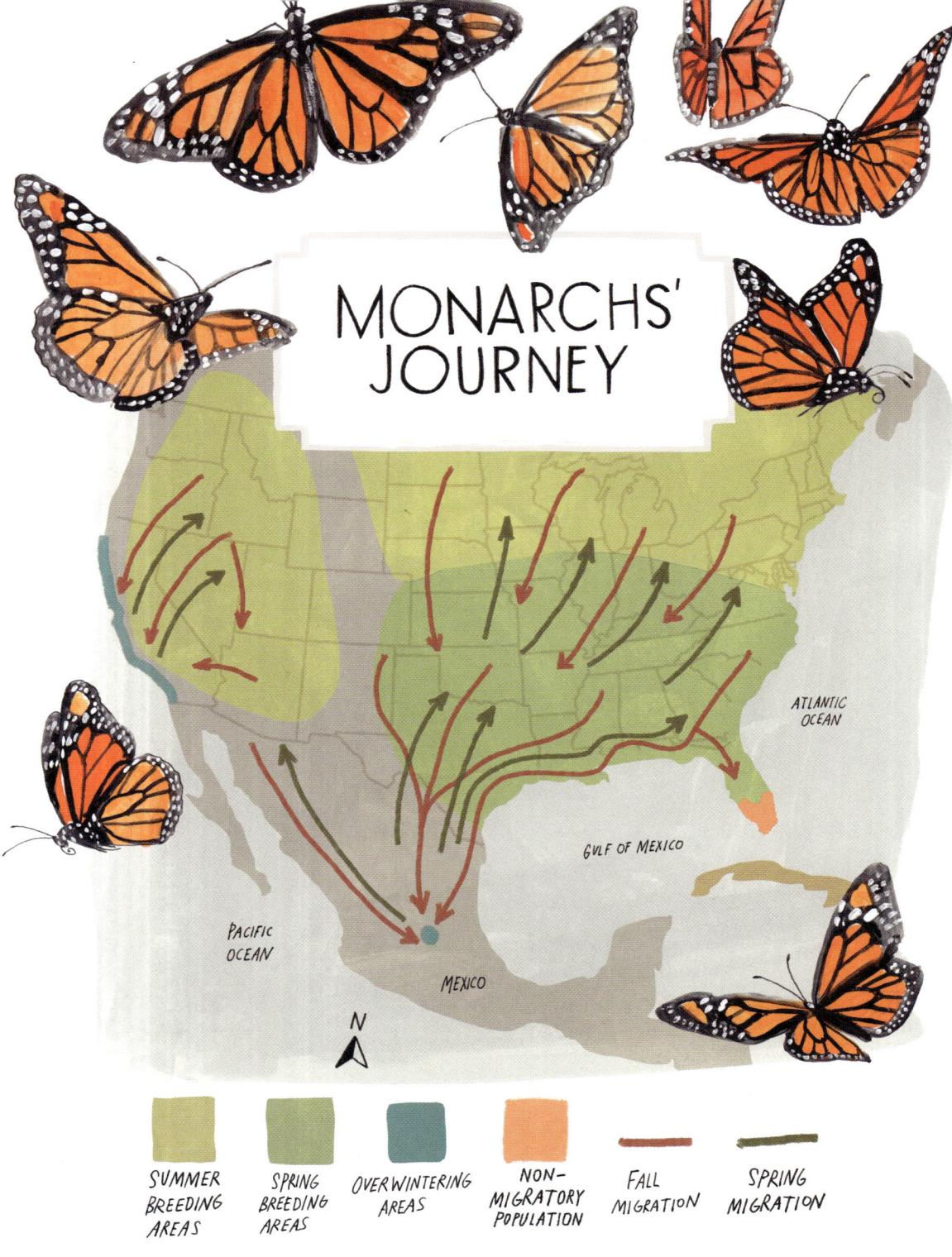

MONARCHS' JOURNEY

ATLANTIC OCEAN

PACIFIC OCEAN

GULF OF MEXICO

MEXICO

N

SUMMER BREEDING AREAS

SPRING BREEDING AREAS

OVERWINTERING AREAS

NON-MIGRATORY POPULATION

FALL MIGRATION

SPRING MIGRATION

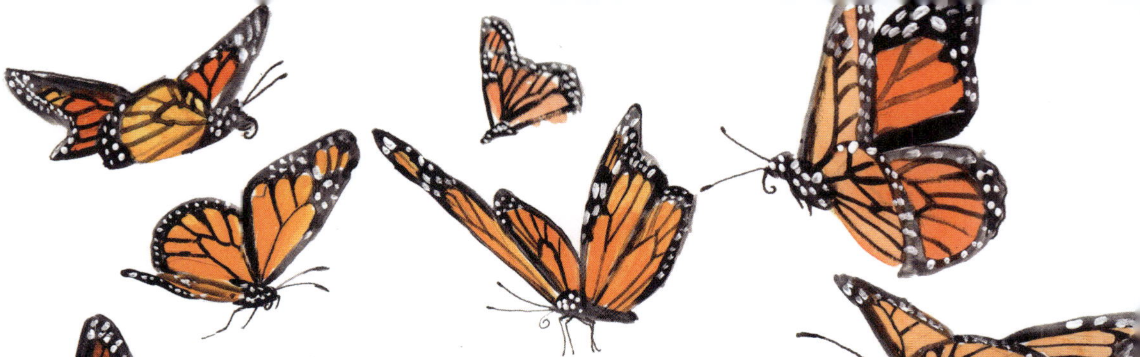

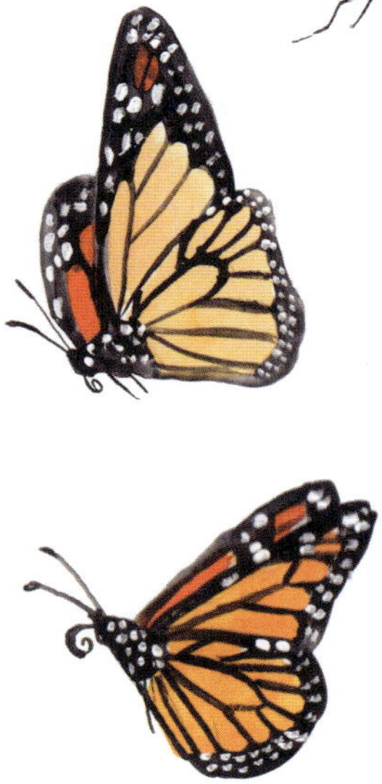

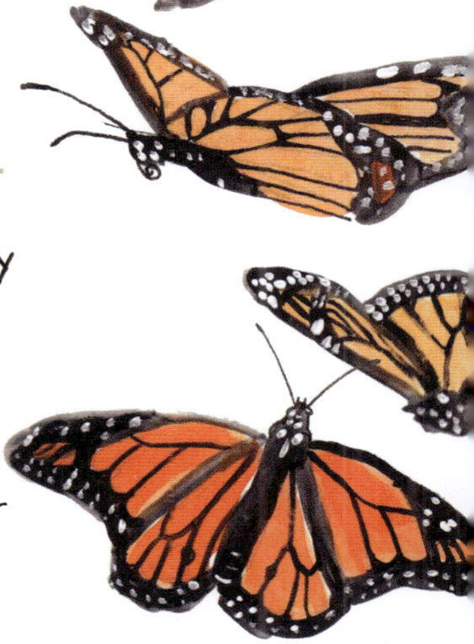

Unlike any other insect, monarch butterflies embark on an incredible journey, traveling nearly 3,000 miles each year from Canada to Central Mexico.

Swaths of monarchs head south in the fall, where they roost in the warm weather until heading back north during the spring. While this behavior somewhat resembles that of a bird's, adult monarch butterflies only live four to five weeks, and thus it takes up to five generations for the monarch to complete the migration from start to finish.

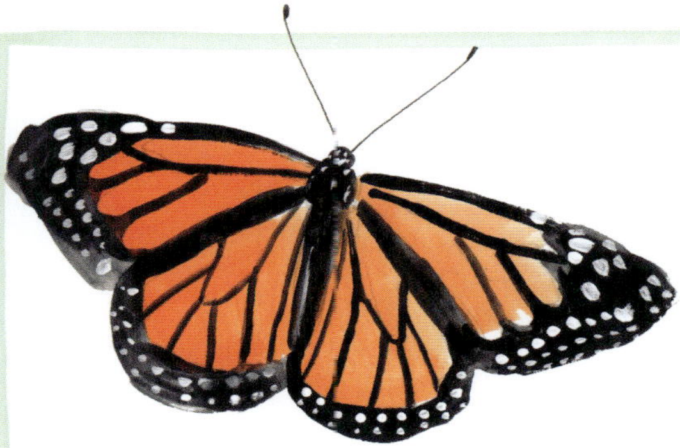

Monarch

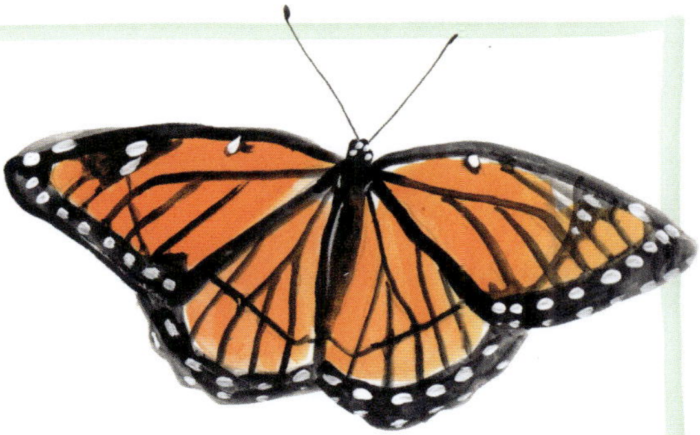

Viceroy

Viceroys evolved to mimic monarchs to avoid predators who think they are as poisonous as monarchs. To tell them apart, look for the horizontal black line running through the Viceroy's hindwings.

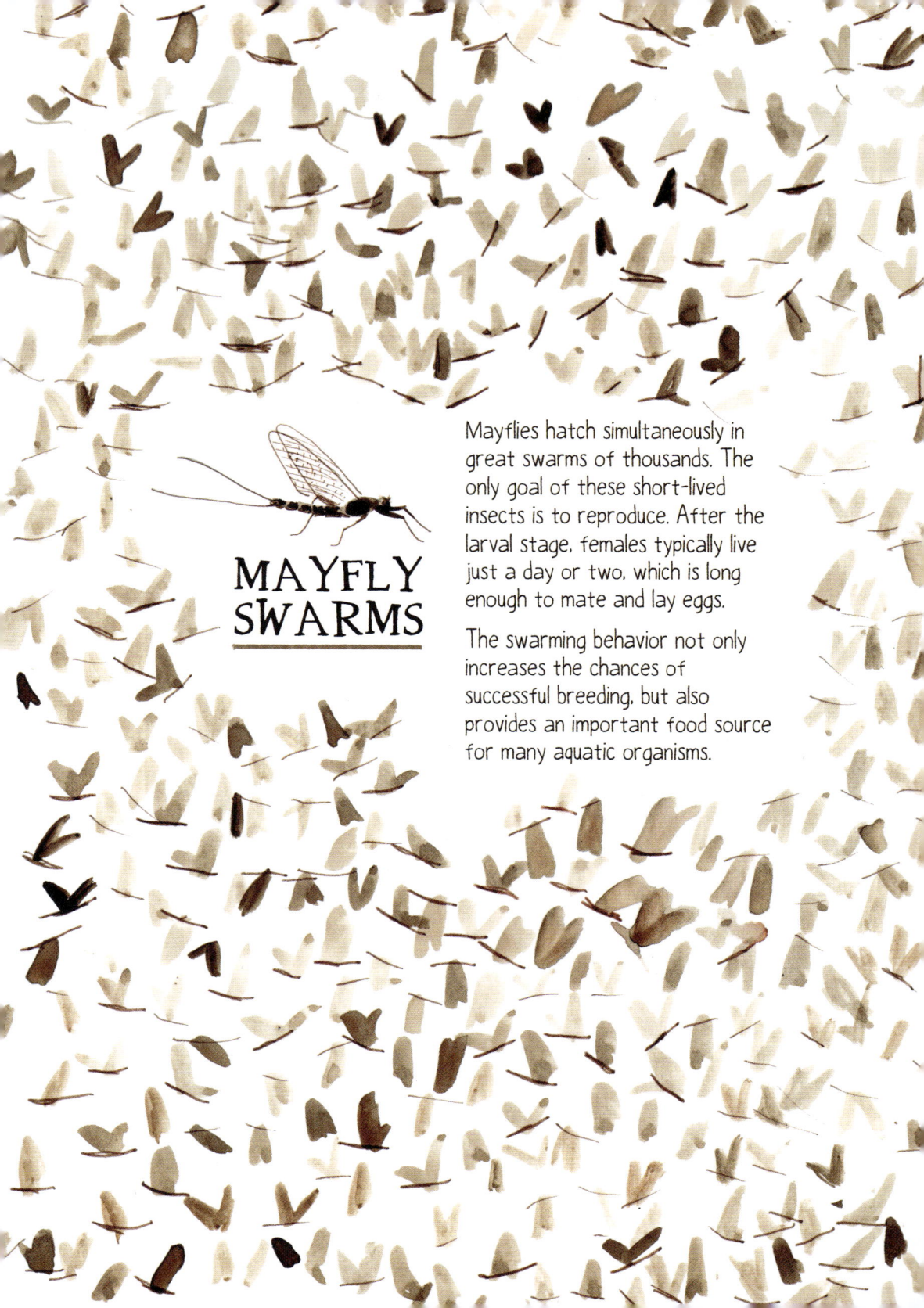

MAYFLY
SWARMS

Mayflies hatch simultaneously in great swarms of thousands. The only goal of these short-lived insects is to reproduce. After the larval stage, females typically live just a day or two, which is long enough to mate and lay eggs.

The swarming behavior not only increases the chances of successful breeding, but also provides an important food source for many aquatic organisms.

INVASION OF THE BODY SNATCHERS

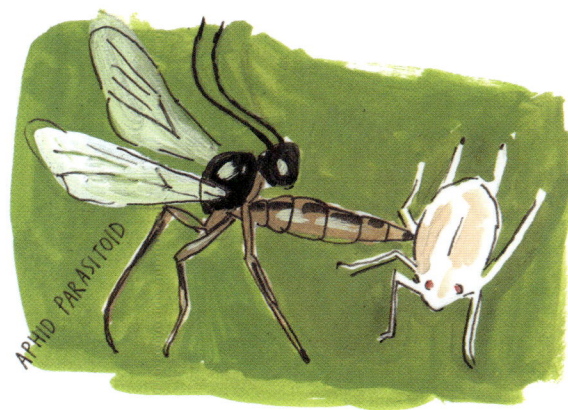

APHID PARASITOID

Parasitoid wasps lay their eggs on or inside of other insects like caterpillars and beetles.

The aphid parasitoid lays eggs inside of aphids. Once the larva develops, it kills the aphid. It turns the aphid into a "mummy" and chews a hole out of the mummy to escape.

Braconid wasps lay eggs in the bodies of hornworm caterpillars. The larvae eventually chew their way out of the caterpillar and spin silk cocoons on the caterpillar's exoskeleton. Adult wasps emerge from these cocoons.

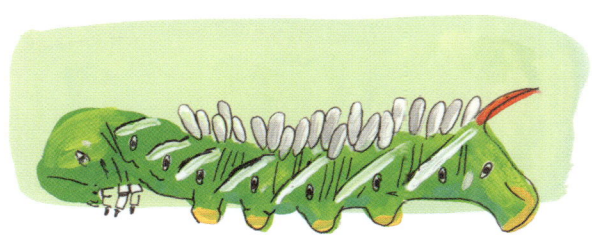

Ichneumon wasps use their long ovipositors to drill through bark and lay eggs on larvae sheltered beneath. The hatchlings eat the paralyzed host, pupate, and chew their way out as adults the following summer.

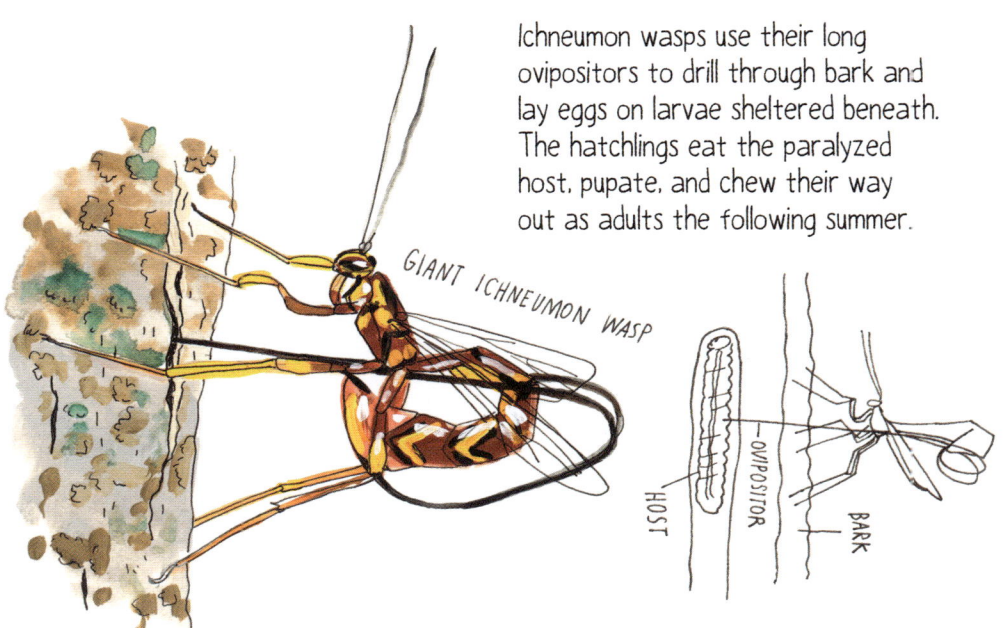

GIANT ICHNEUMON WASP

HOST

OVIPOSITOR

BARK

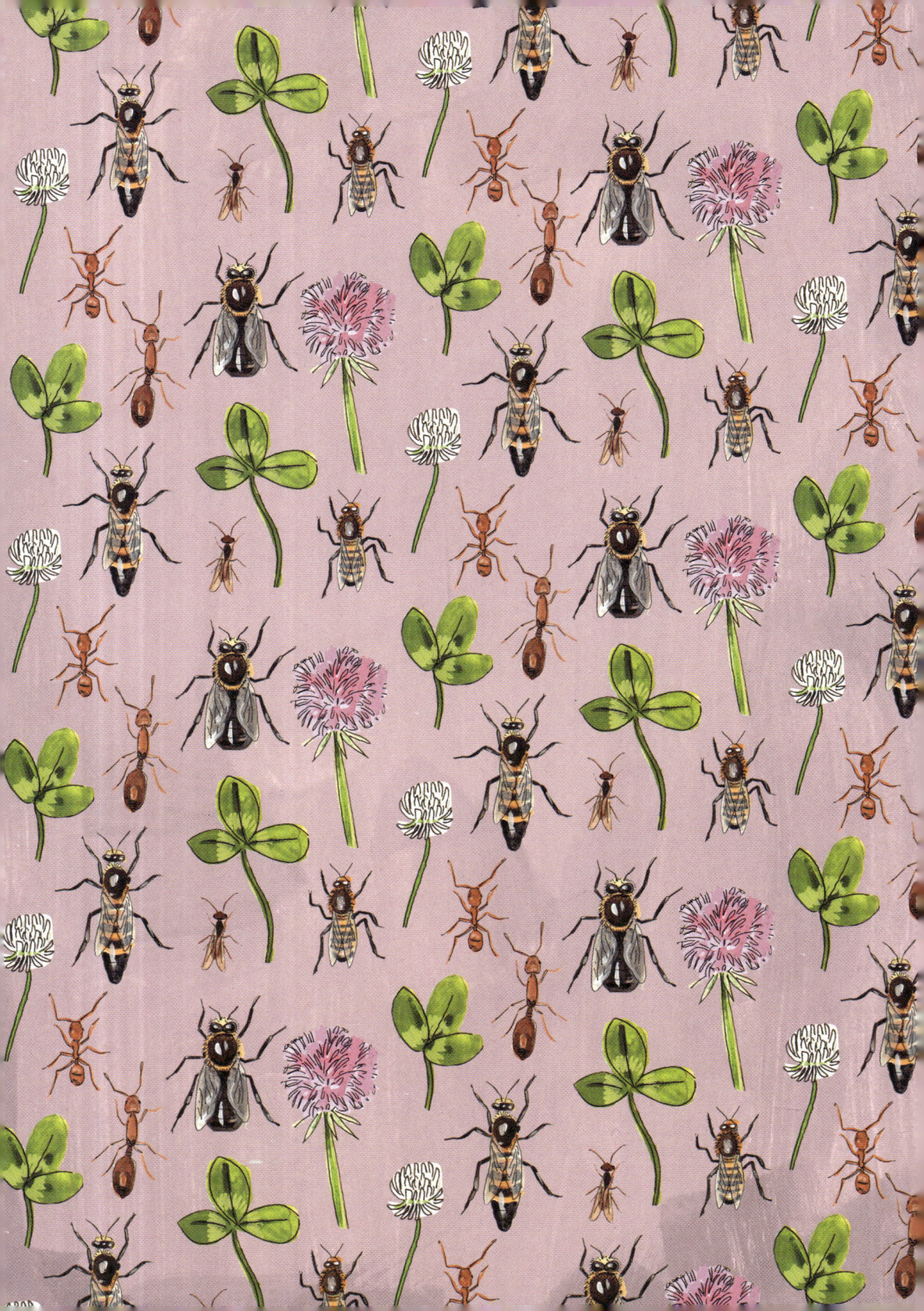

CHAPTER 3

Hive Minded

EUSOCIAL

Eusociality is a cooperative social system found among ants, bees, wasps, and termites. It is marked by the division of labor into reproductive and non-reproductive groups, overlapping generations, and cooperative brood care.

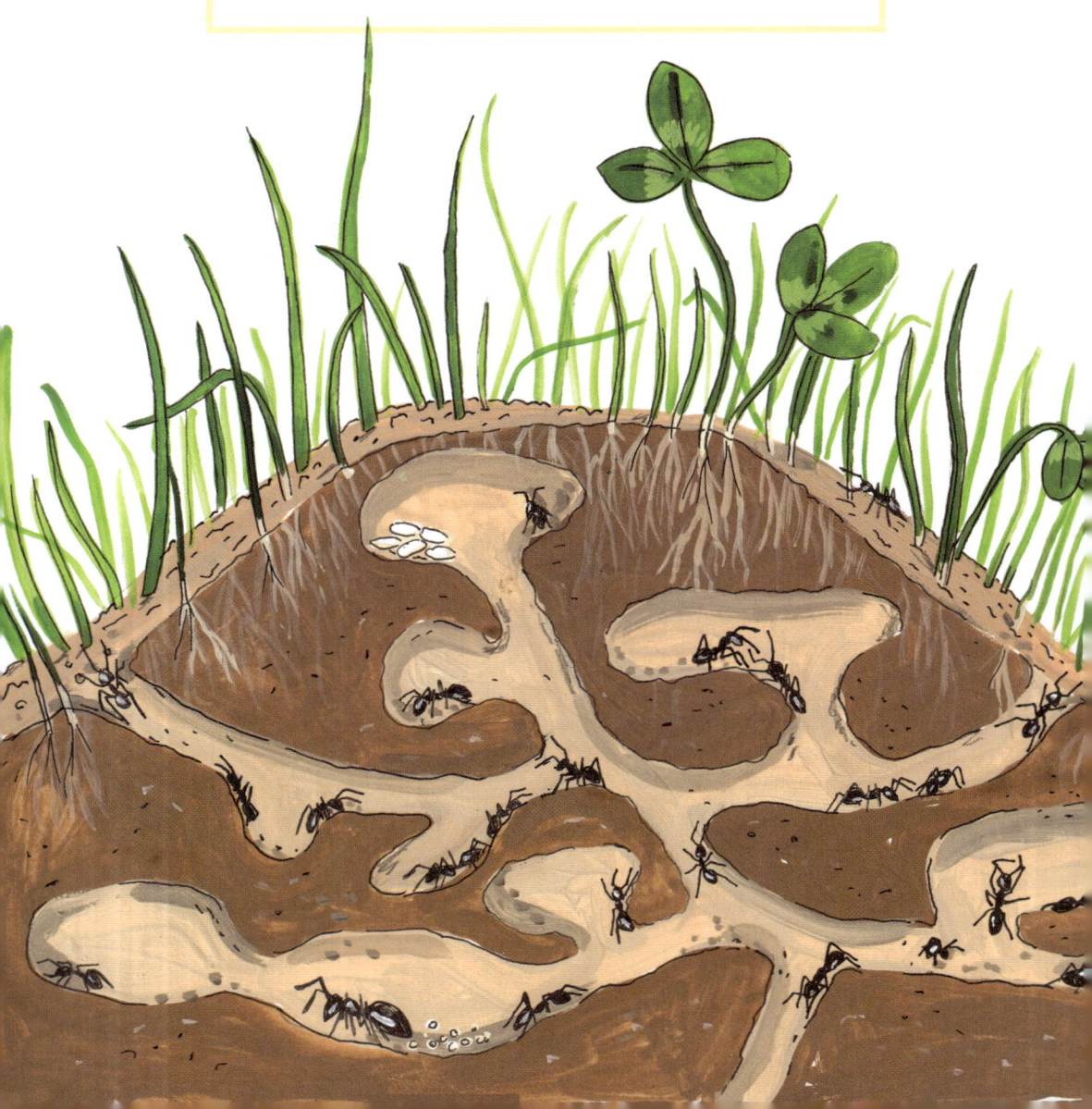

ABOUT ANTS

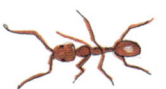

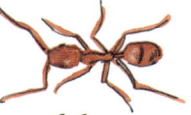

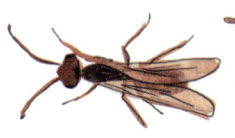

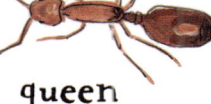

worker

These sterile female ants forage for food, feed the larvae, build and defend the nest, and keep the nest clean by taking out the trash.

soldier

Slightly larger than workers, soldier ants are also sterile females. They protect the colony from predators and clear pathways for the worker ants to reach food.

drone

The sole purpose of these winged males is to mate with the queen, after which they die.

queen

The matriarch has one job, to lay eggs. She can produce 1,000 eggs a day and live up to 30 years.

There are about 16,000 known species of ants, and they are found on every continent on Earth except Antarctica.

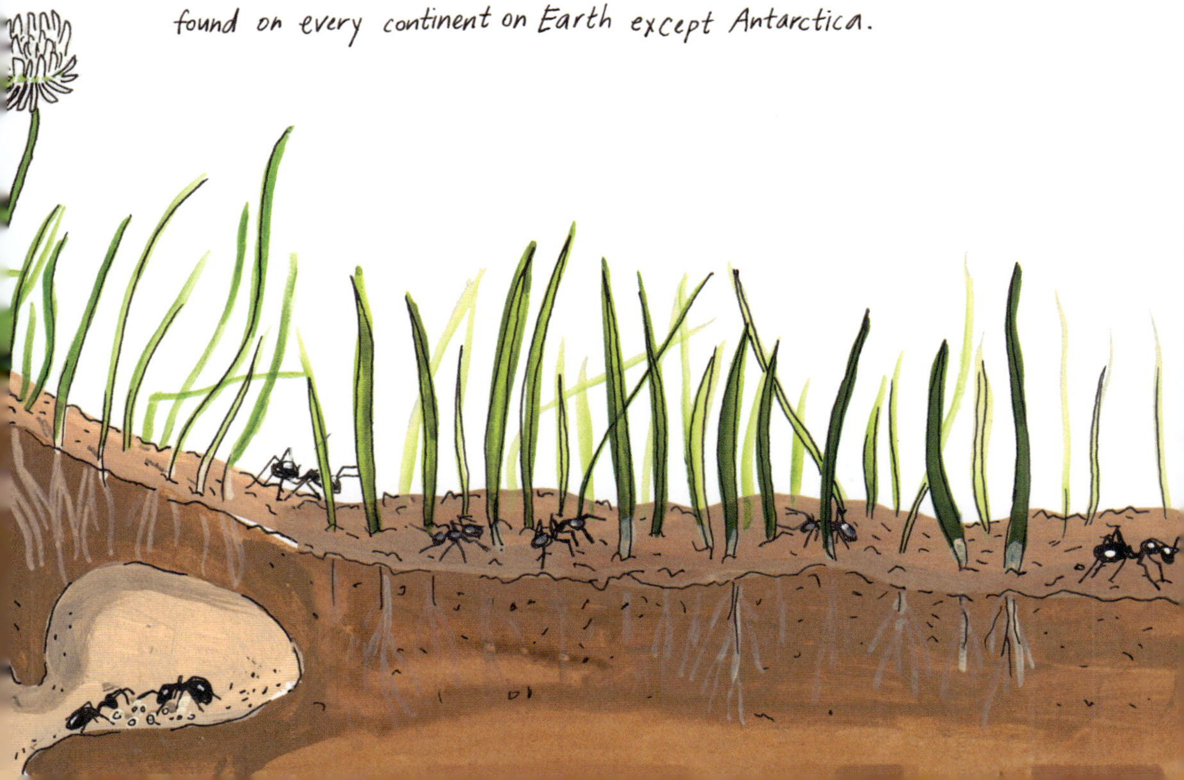

ANATOMY OF AN ANT

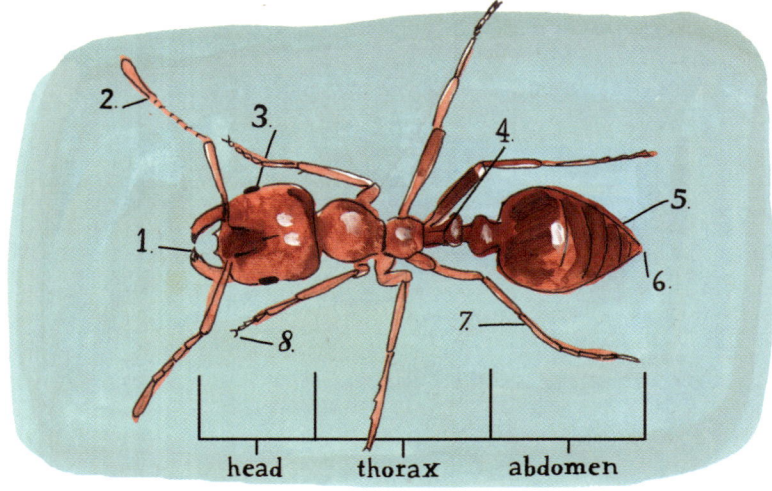

head   thorax   abdomen

1. **mandibles** Operating in a sideways fashion, ants use their mandibles (jaws) for many tasks, including cutting, carrying, and trapping.

2. **antennae** Antennae are used to help recognize mates and detect enemies. If an ant loses its antennae, it becomes unable to communicate with its colony.

3. **compound eyes** Ants have two large compound eyes, and see one or two feet in front of themselves. Some species are completely blind. Ants use their antennae far more than their eyes.

4. **pedicel** The narrow waist connecting the thorax and abdomen is a key characteristic of ants.

5. **gaster** The bulbous part of an ant's abdomen contains the ant's stomach, vital organs, and reproductive parts.

6. **stinger** Some ants can sting. While the sting from a fire ant is considered mild, the bullet ant's sting has been described as "like walking over flaming charcoal with a three-inch nail embedded in your heel."

7. **legs** Ants have sensory organs on their legs to help detect vibrations, sounds, and smells.

8. **claws** Ants have two hooked claws on each leg to help grip while climbing.

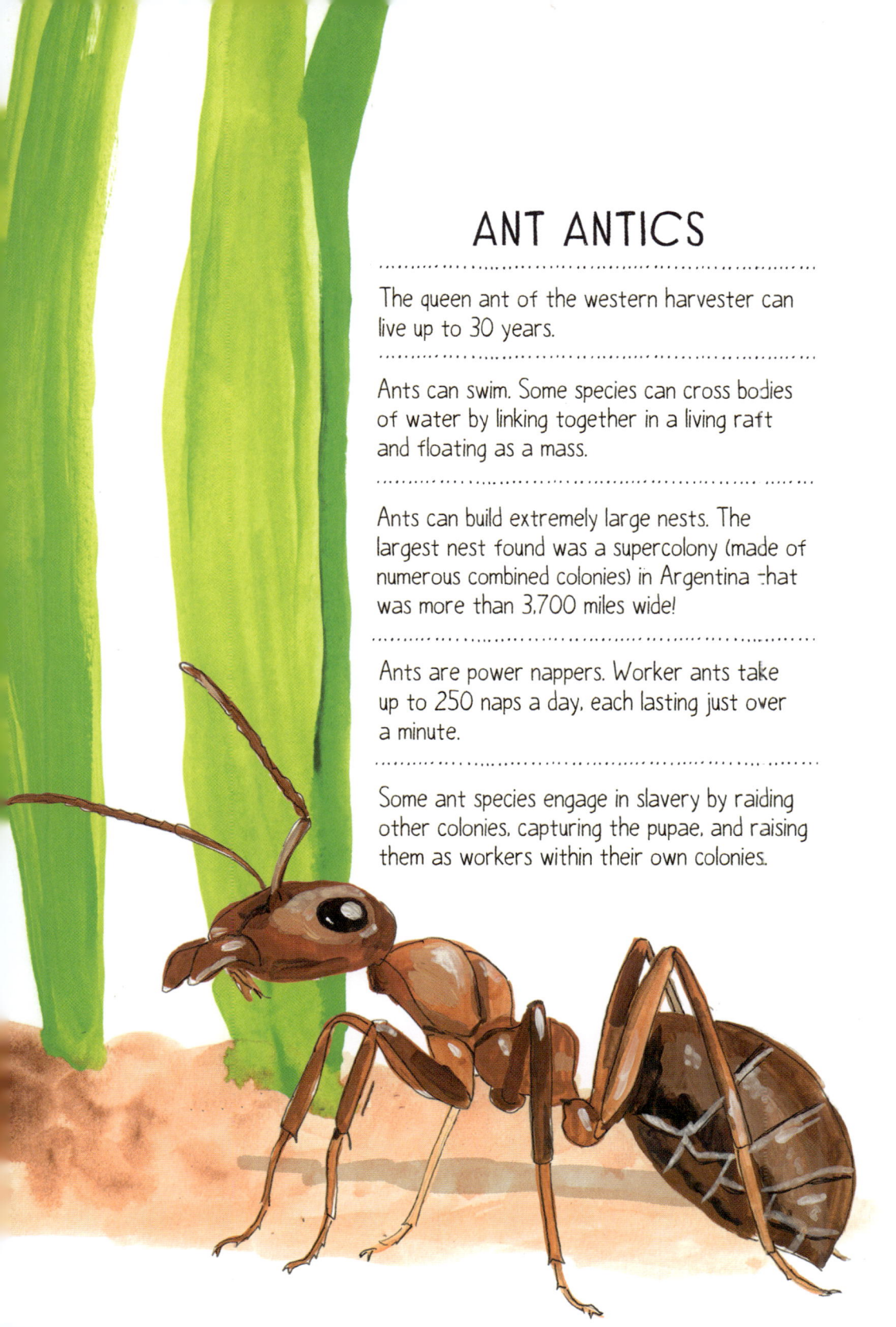

ANT ANTICS

The queen ant of the western harvester can live up to 30 years.

Ants can swim. Some species can cross bodies of water by linking together in a living raft and floating as a mass.

Ants can build extremely large nests. The largest nest found was a supercolony (made of numerous combined colonies) in Argentina that was more than 3,700 miles wide!

Ants are power nappers. Worker ants take up to 250 naps a day, each lasting just over a minute.

Some ant species engage in slavery by raiding other colonies, capturing the pupae, and raising them as workers within their own colonies.

ANT SPECIES

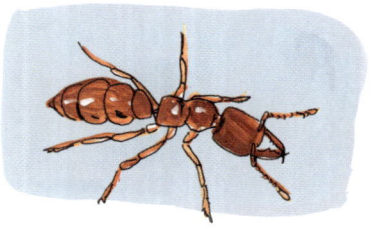

The DRACULA ant is an endangered species, native to Madagascar. It is named for its grisly habit of drinking the blood of its young.

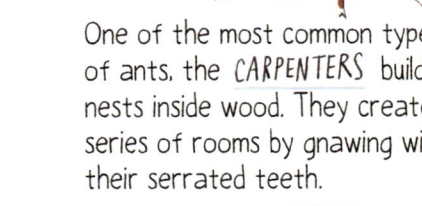

One of the most common types of ants, the CARPENTERS build nests inside wood. They create a series of rooms by gnawing with their serrated teeth.

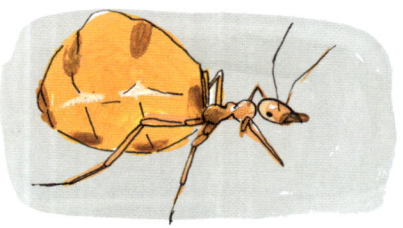

HONEYPOT worker ants store food in their enlarged abdomens, which in turn is used to nourish other ants in the colony.

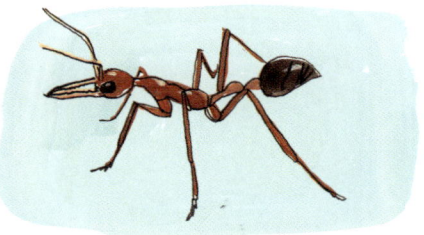

BULLDOG, or "jack jumper," ants can be identified by their elongated mandibles. They are known to be extremely aggressive, and may jump when agitated.

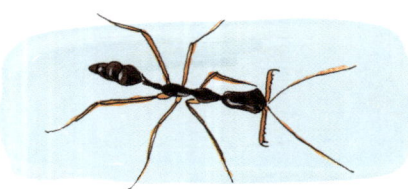

Found in Central and South America, the TRAP-JAW ant snaps its powerful mandibles shut at speeds up to 145 mph.

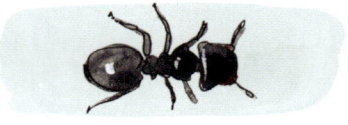

The TURTLE or door head ant has been observed using its oddly shaped head to "plug" entranceways shut, keeping invaders out of their nesting tunnels.

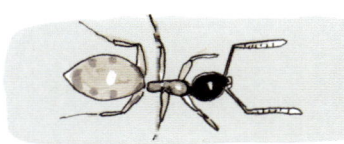

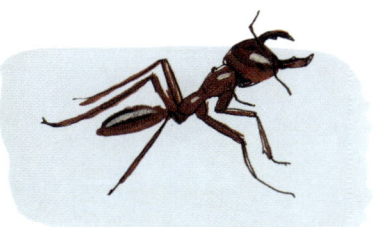

Tiny GHOST ants have a dark head and a nearly translucent abdomen and legs. When running on light surfaces, the body and legs appear to vanish, leaving only the head and thorax visible.

Found in Africa, blind SIAFU ants hunt together in large armies that are capable of completely devouring small animals.

Like some other ant species, soldier siafu ants form bridges with their collective bodies so that they can cross gaps during a hunt.

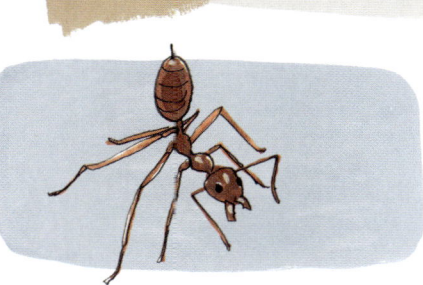

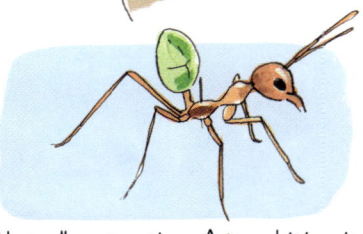

Aggressive stinging FIRE ants live in mounds in large colonies that can have upwards of 200,000 members.

Not all ants sting. After biting its victim, the GREEN TREE ant swings its tail under to squirt formic acid into the wound, which causes a burning sensation and deters further attacks.

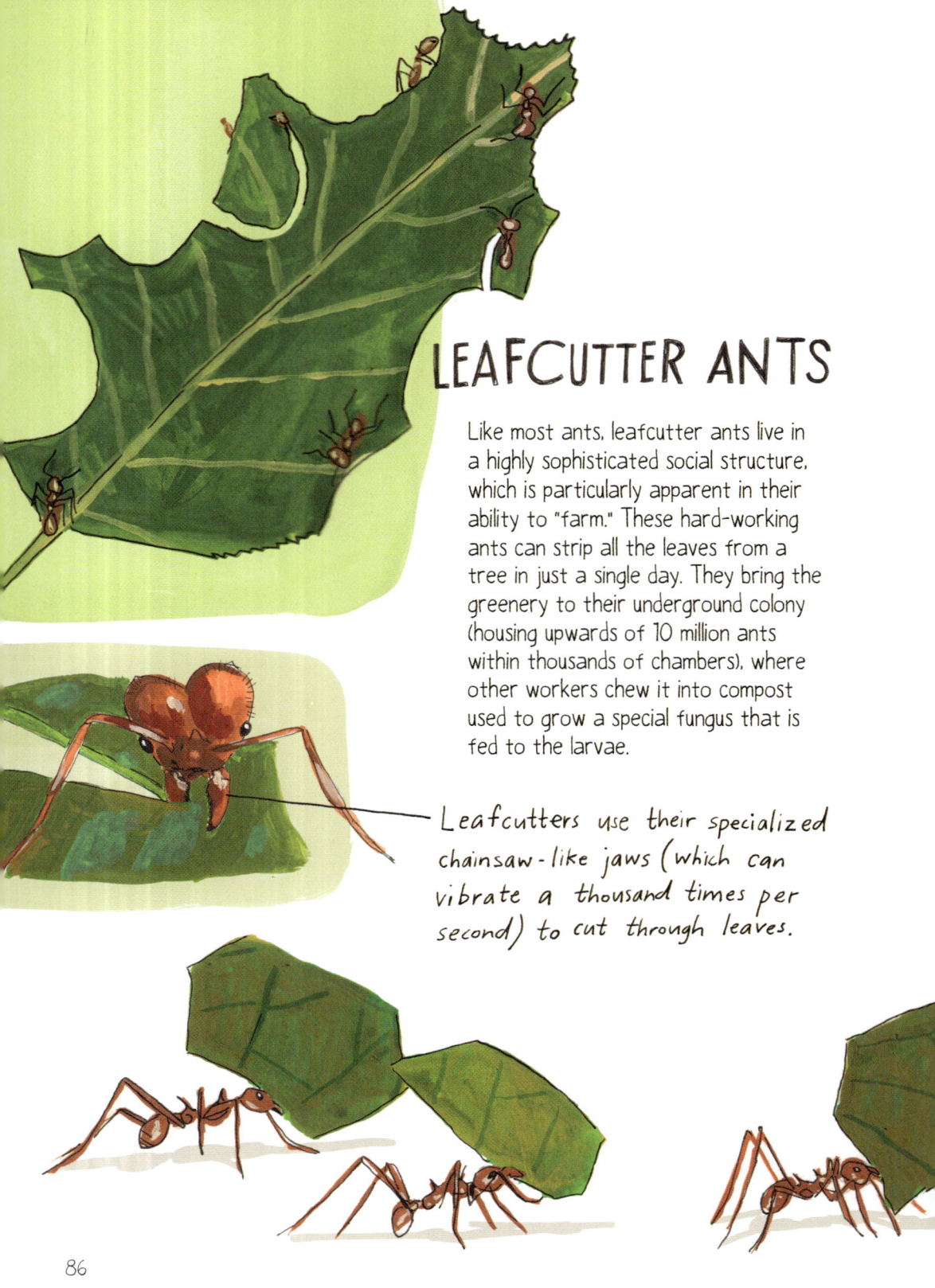

LEAFCUTTER ANTS

Like most ants, leafcutter ants live in a highly sophisticated social structure, which is particularly apparent in their ability to "farm." These hard-working ants can strip all the leaves from a tree in just a single day. They bring the greenery to their underground colony (housing upwards of 10 million ants within thousands of chambers), where other workers chew it into compost used to grow a special fungus that is fed to the larvae.

Leafcutters use their specialized chainsaw-like jaws (which can vibrate a thousand times per second) to cut through leaves.

Smaller worker ants do the work of chewing the leaves into paste pellets. The pellets are fed to a bed of fungus, which in turn creates even more fungus. This farming process requires very careful tending to keep the fungus healthy.

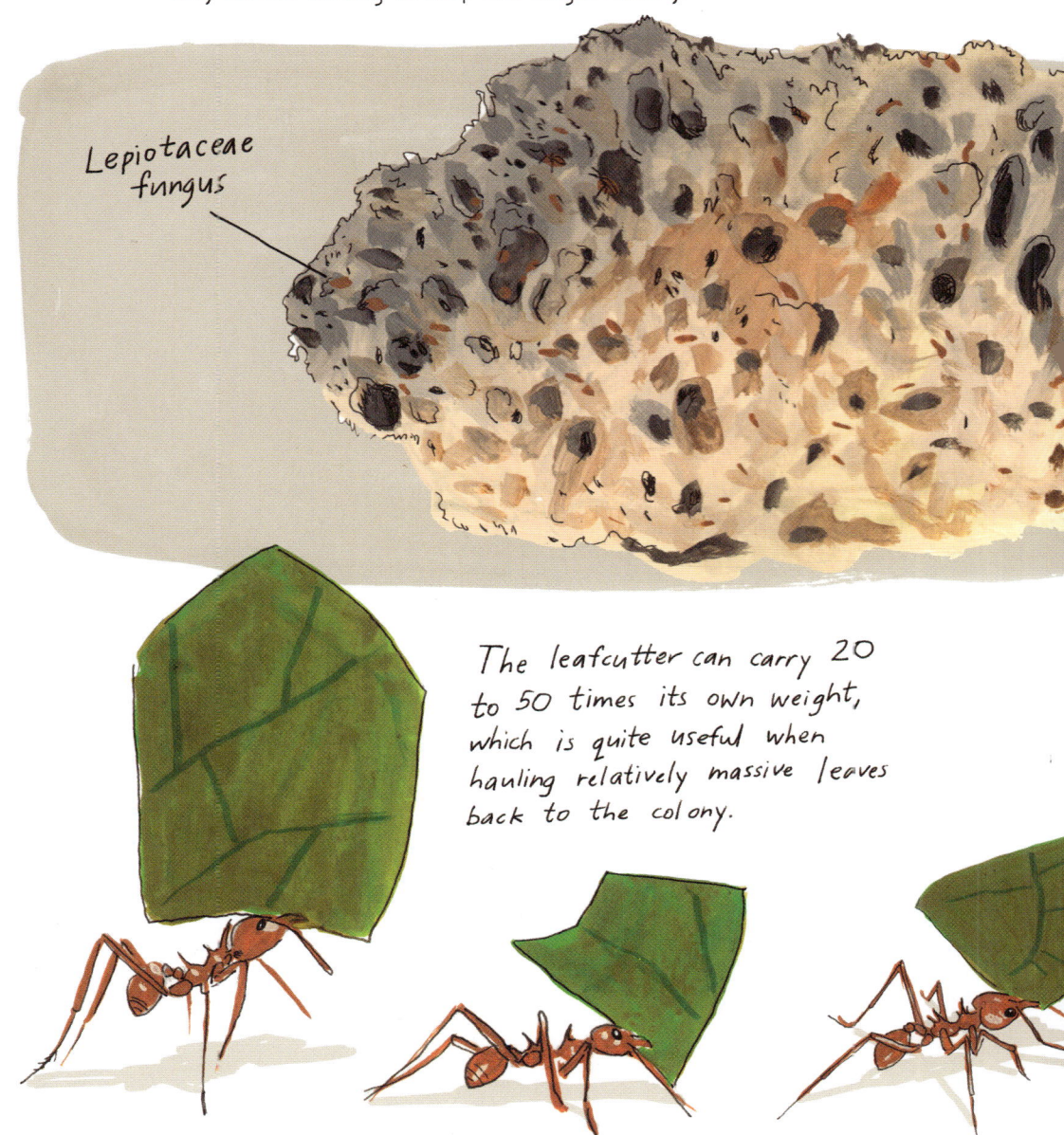

Lepiotaceae fungus

The leafcutter can carry 20 to 50 times its own weight, which is quite useful when hauling relatively massive leaves back to the colony.

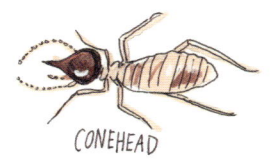
CONEHEAD

TERMITES

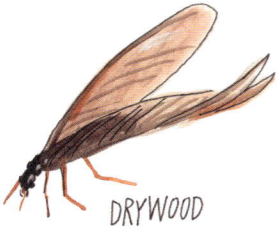
DRYWOOD

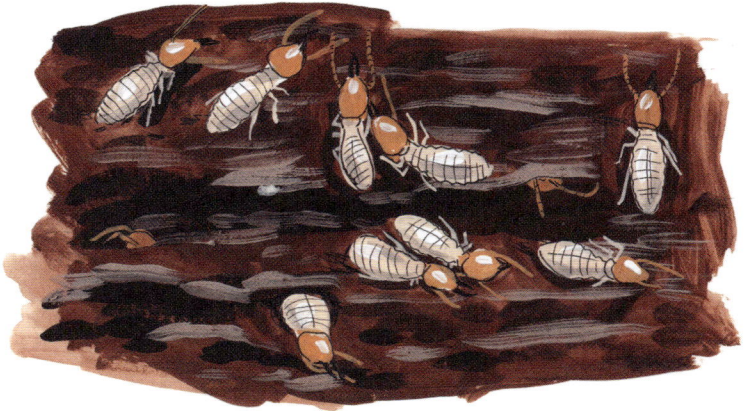

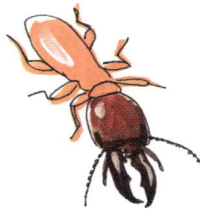
DAMPWOOD

Termites are eusocial insects belonging to the order Blattodea. While termites are commonly mistaken as a type of ant, they are much more closely related to cockroaches and mantids. Termites have long been considered pests to humans because of their appetite for cellulite, which is found in wood, and thus their ability to destroy the structure of a home and the furniture in it.

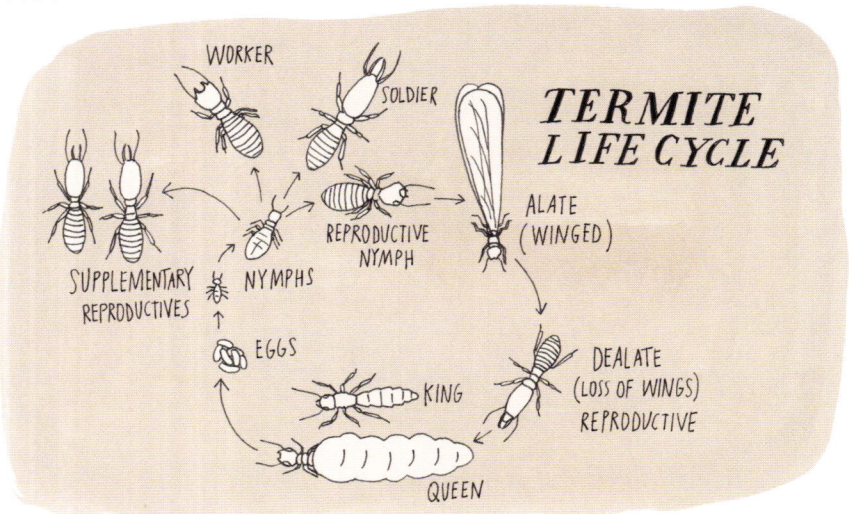

WORKER

SOLDIER

TERMITE LIFE CYCLE

REPRODUCTIVE NYMPH

ALATE (WINGED)

SUPPLEMENTARY REPRODUCTIVES

NYMPHS

EGGS

KING

DEALATE (LOSS OF WINGS) REPRODUCTIVE

QUEEN

While termites have social structures similar to bees, wasps, and ants, there is one distinct difference: the termite king. The king lives in the colony and fertilizes all the eggs, unlike drone bees. The queen's long abdomen houses her ovaries, which pump out eggs—some species can lay 30,000 eggs a day!

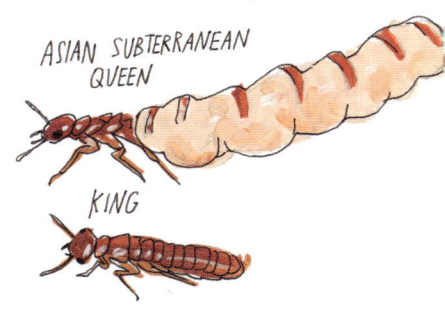

ASIAN SUBTERRANEAN QUEEN

KING

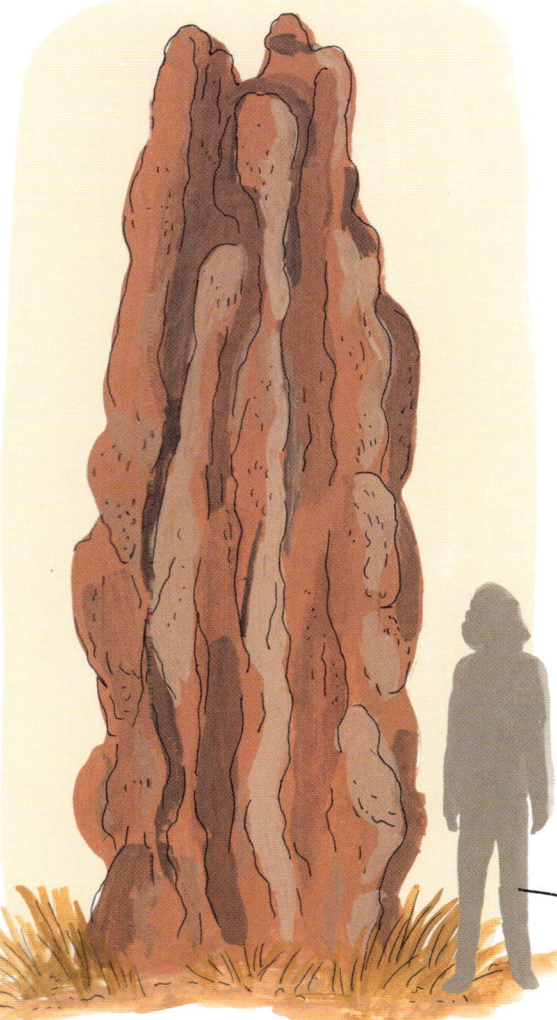

Mound-building termites are found in South America, Australia, and Africa. Their homes are made of soil, termite saliva, and dung.

Cathedral termites from the Northern Territory of Australia make towering mounds that are more than 15 feet tall. The conical shape of the mound allows hot air to circulate upward and cool the entire structure.

HUMAN FOR SCALE

ANATOMY OF A BEE

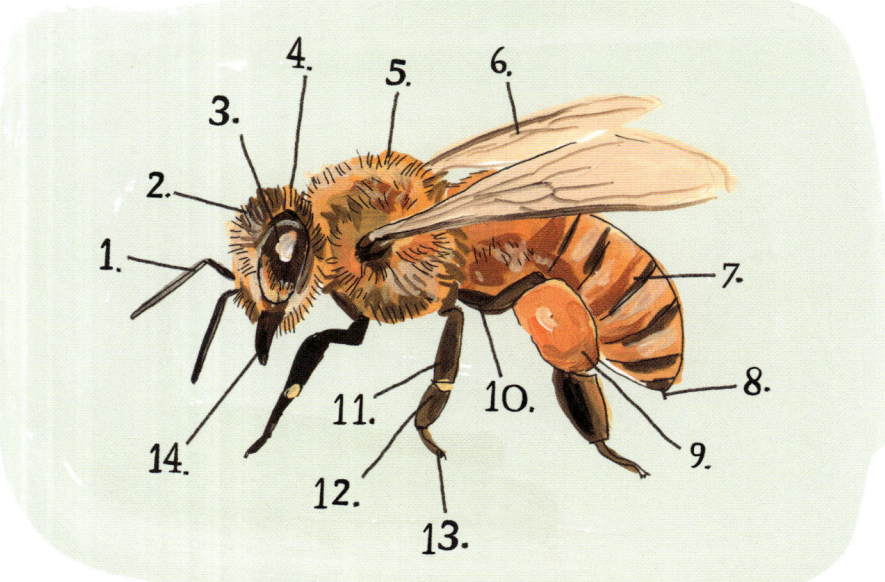

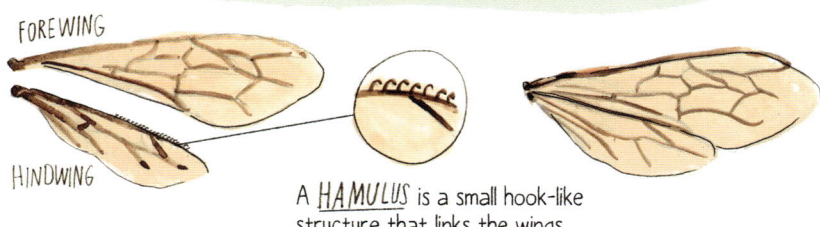

FOREWING

HINDWING

A *HAMULUS* is a small hook-like structure that links the wings.

1. antennae
2. 3 simple eyes (on forehead)
3. compound eye
4. head
5. thorax
6. wings
7. abdomen
8. stinger
9. pollen basket
10. femur
11. tibia
12. tarsus
13. tarsal claw
14. mandible

INSIDE A BEE

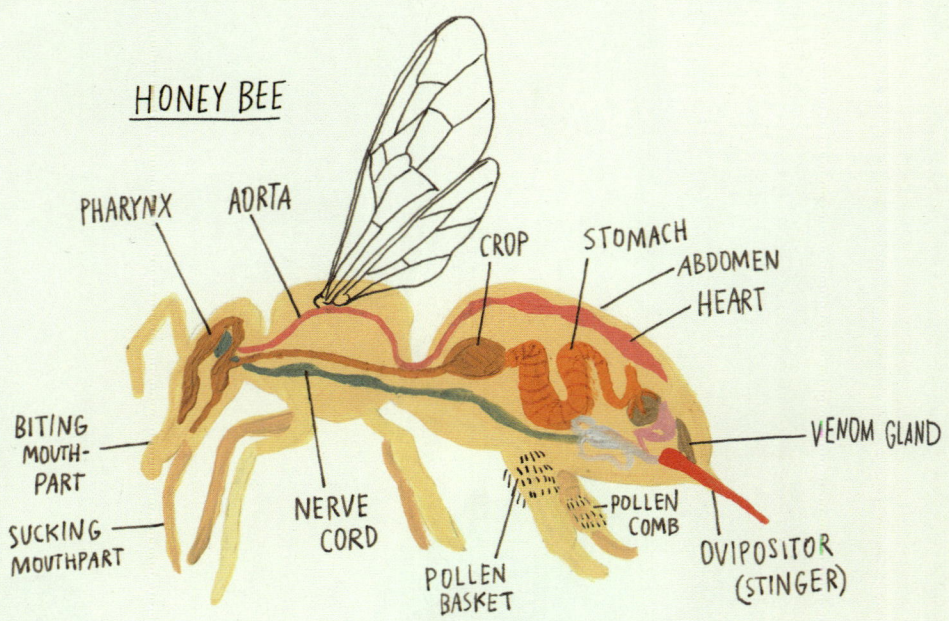

HONEY BEE

PHARYNX · AORTA · CROP · STOMACH · ABDOMEN · HEART

BITING MOUTH-PART · SUCKING MOUTHPART · NERVE CORD · POLLEN BASKET · POLLEN COMB · OVIPOSITOR (STINGER) · VENOM GLAND

The PHARYNX is the muscular area that draws in food, like nectar.

The CROP (honey sack) is used to store and transport liquid food.

The NERVE CORD coordinates signals from the brain to various body parts.

The AORTA acts as a long heart, circulating blood to the various organs.

The STOMACH digests food and absorbs nutrients. The hindgut is where certain salts and amino acids are absorbed before excretion of feces.

The VENOM GLAND stores and delivers venom to the stinger.

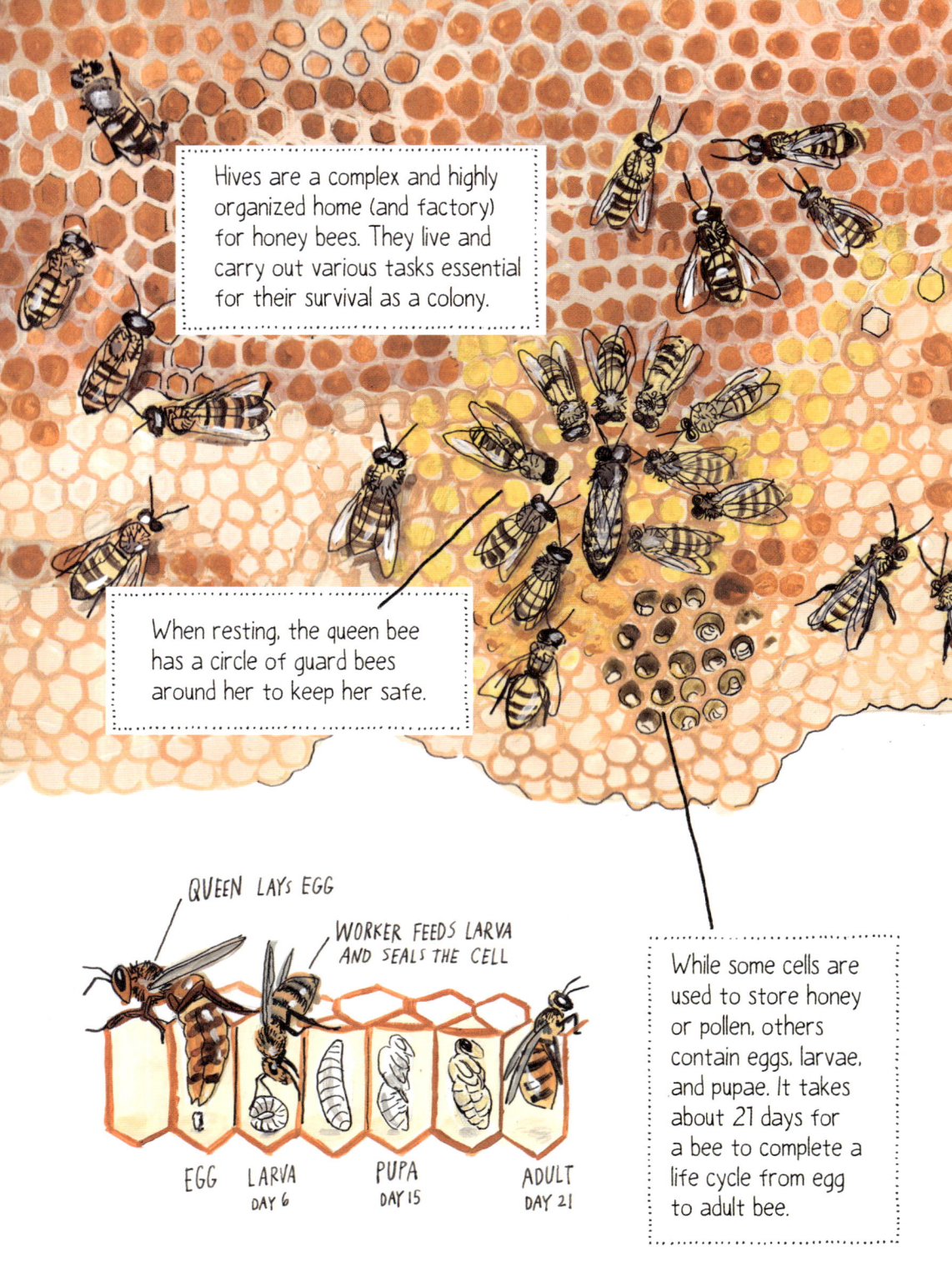

Hives are a complex and highly organized home (and factory) for honey bees. They live and carry out various tasks essential for their survival as a colony.

When resting, the queen bee has a circle of guard bees around her to keep her safe.

QUEEN LAYS EGG

WORKER FEEDS LARVA AND SEALS THE CELL

While some cells are used to store honey or pollen, others contain eggs, larvae, and pupae. It takes about 21 days for a bee to complete a life cycle from egg to adult bee.

EGG   LARVA   PUPA   ADULT
       DAY 6   DAY 15 DAY 21

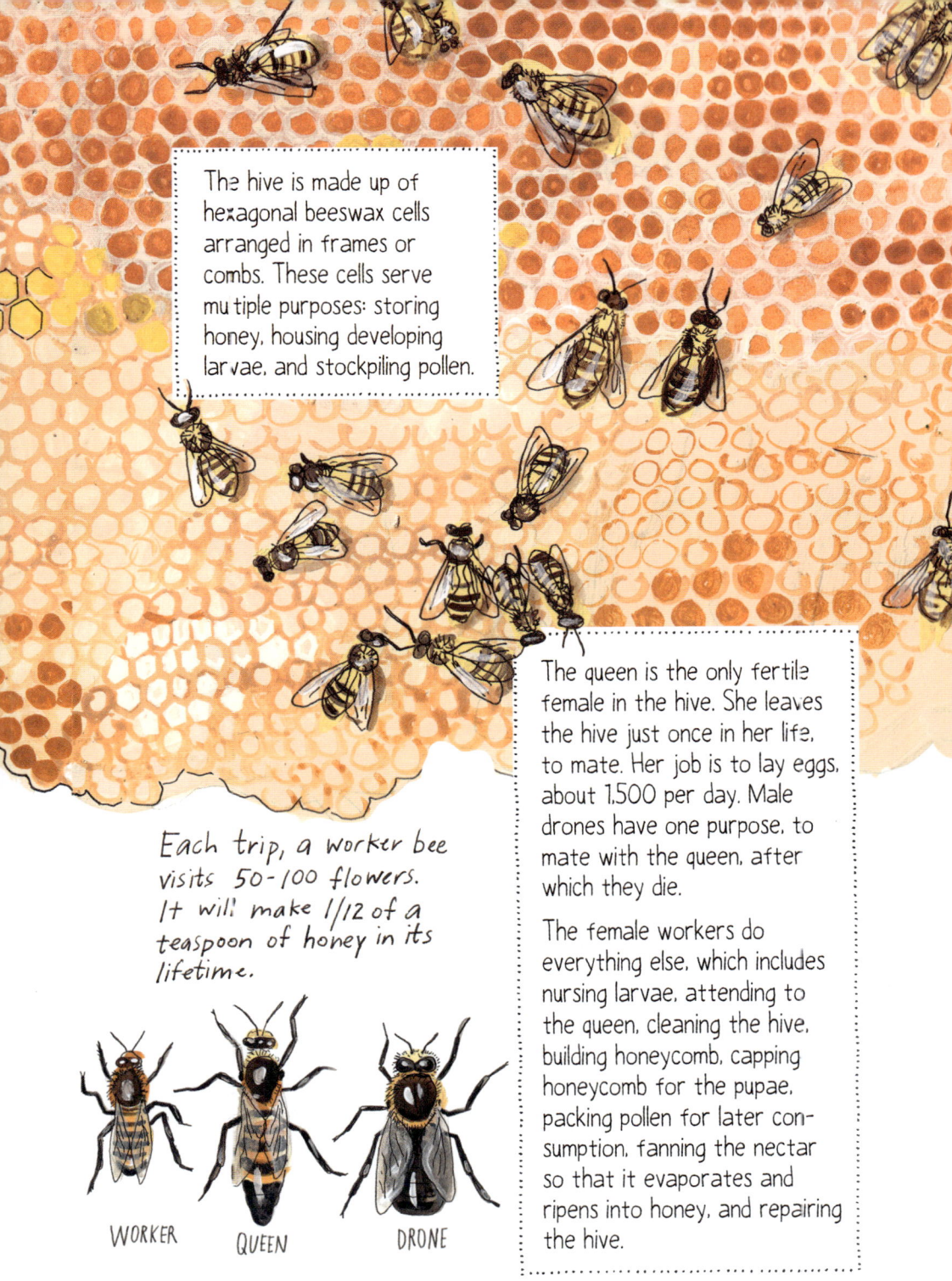

The hive is made up of hexagonal beeswax cells arranged in frames or combs. These cells serve multiple purposes: storing honey, housing developing larvae, and stockpiling pollen.

The queen is the only fertile female in the hive. She leaves the hive just once in her life, to mate. Her job is to lay eggs, about 1,500 per day. Male drones have one purpose, to mate with the queen, after which they die.

The female workers do everything else, which includes nursing larvae, attending to the queen, cleaning the hive, building honeycomb, capping honeycomb for the pupae, packing pollen for later consumption, fanning the nectar so that it evaporates and ripens into honey, and repairing the hive.

Each trip, a worker bee visits 50-100 flowers. It will make 1/12 of a teaspoon of honey in its lifetime.

WORKER   QUEEN   DRONE

BEE SPECIES

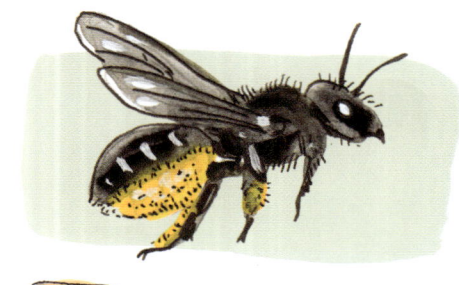

MASON

Just 250 to 300 of these highly efficient fruit tree pollinators can pollinate an entire acre of apple trees. While most bees carry pollen on their legs, the mason bee has a special pollen-carrying structure (scopa) on its abdomen.

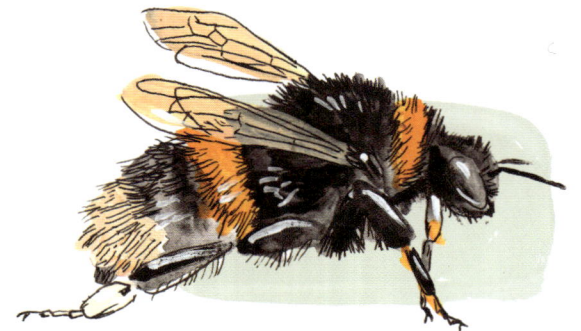

BUMBLE

Most of the 250+ species of bumble bees live socially in colonies with a single queen. Unlike honey bees, bumble bees can sting multiple times, although they are much less likely to sting.

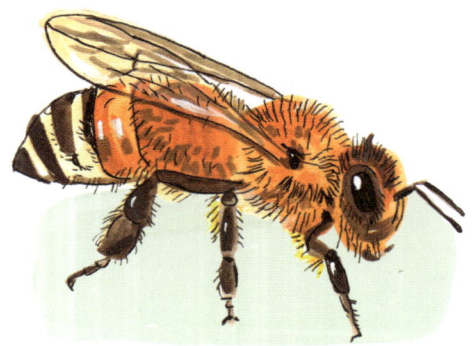

HONEY

These hard-working bees produce two to three times more honey than they need, allowing us to enjoy some of their product. They will sting to defend their territory, but typically die after losing their stinger.

GREEN SWEAT

Named for its attraction to the salt found in sweat, this colorful bee emerges in the spring and feeds on the nectar of flowers as well as the honeydew of aphid colonies.

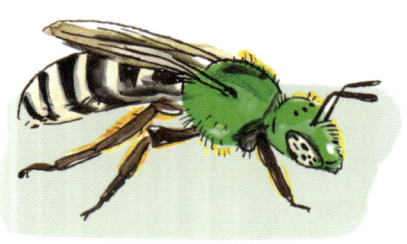

DIGGER

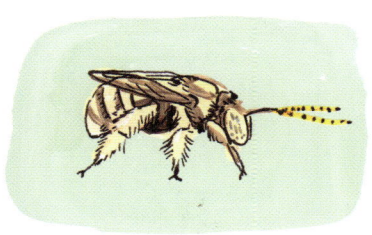

These robust, hairy bees are easy to recognize by their visibly protruding faces and their disproportionately short wings that create a high-pitched buzz while hovering above plants. The digger bee builds nests in the ground by digging with its front legs and using its mouth to soften the dirt.

SMALL CARPENTER

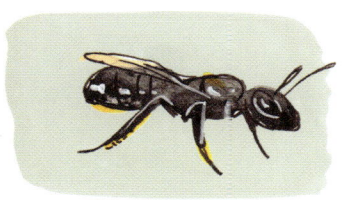

These little bees (distantly related to the larger and more familiar carpenter bee) make nests in dried stems and sticks, carving out individual chambers for each egg.

FAIRY

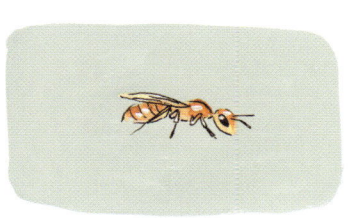

There are hundreds of species of fairy bees, with *Perdita minima* being one of the smallest in the world at slightly less than 2 mm long. This non-stinging bee lives in extremely small nests within sandy soils. Although tiny, it can still carry quite a bit of pollen on its hairy legs.

MINING

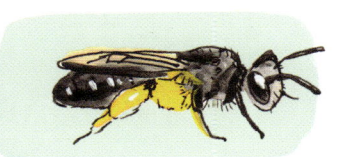

There are more than 1,500 species of mining bee. Instead of building a communal hive, each mining bee digs its own nest in an underground hole.

LEAFCUTTER

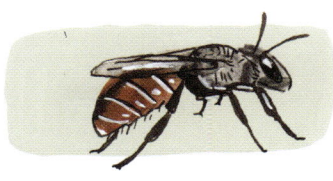

These important pollinators rarely sting. They get their name from the leaves that they cut to build nests in rotting wood.

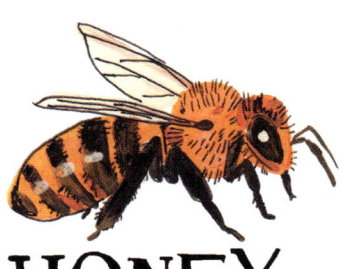

HONEY BEES

VS.

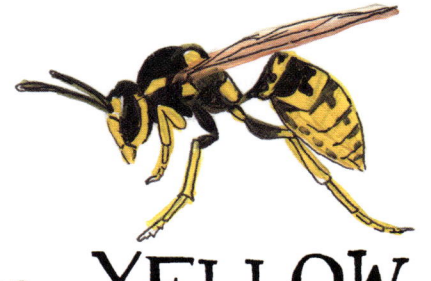

YELLOW JACKETS

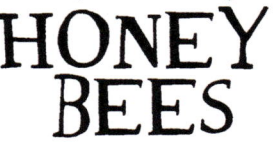

HONEY BEES	YELLOW JACKETS
FUZZY BODY	SMOOTH BODY
EAT NECTAR + POLLEN	EAT INSECTS
MAKE HONEY	DON'T MAKE HONEY
RARELY STING	STING IF AGGRAVATED
GENTLE IN NATURE	DEFENSIVE
SUBSTANTIAL POLLINATION CAPACITY	MINIMAL POLLINATION CAPACITY
CAN ONLY STING ONCE	CAN STING REPEATEDLY

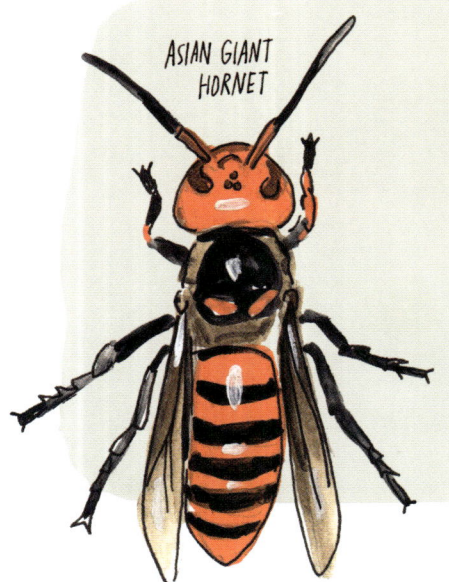

ASIAN GIANT HORNET

AND A HORNET?

A HORNET IS A TYPE OF WASP, GENERALLY WITH BIGGER HEADS AND ABDOMENS.

Also known as the "murder hornet," the Asian giant hornet can wipe out a honey bee nest in just hours, decapitating the entire colony and flying off with the bodies to feed their young.

WASP NESTS

Some wasps build their nests underground, while others hang their nests from trees or hide them in hollow crevices. Like beehives, wasp nests are also built with a series of connecting hexagonal cells; however, the walls are made from chewed wood, clay, or mud.

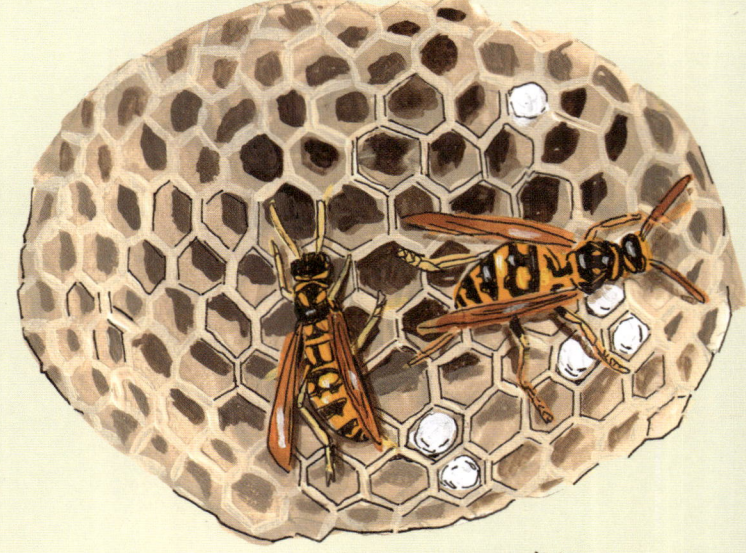

PAPER WASP NEST

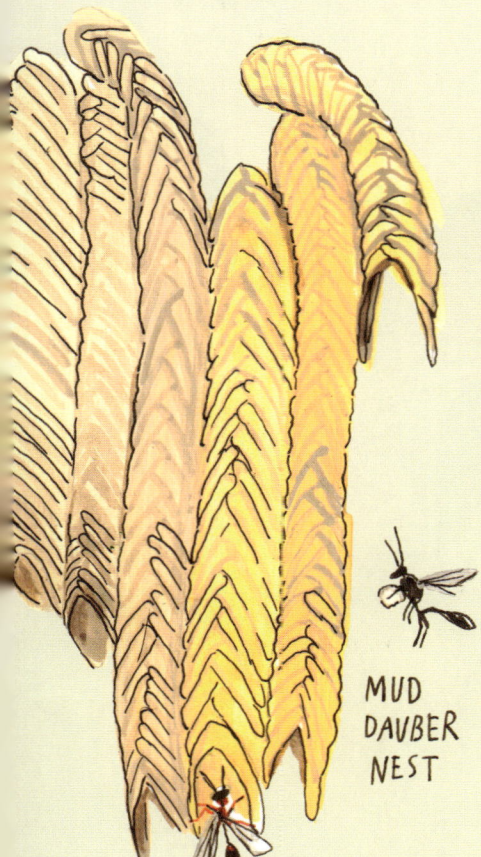

MUD DAUBER NEST

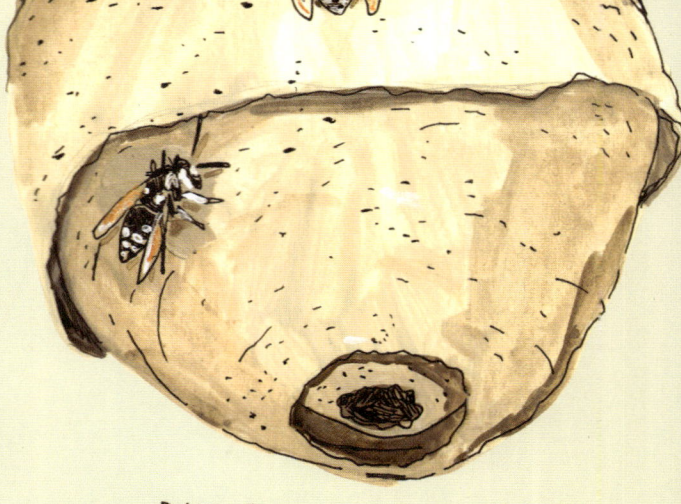

BALD-FACED HORNET NEST

WASP SPECIES

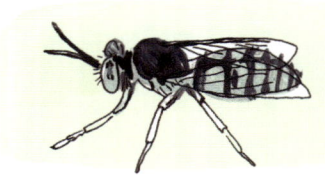

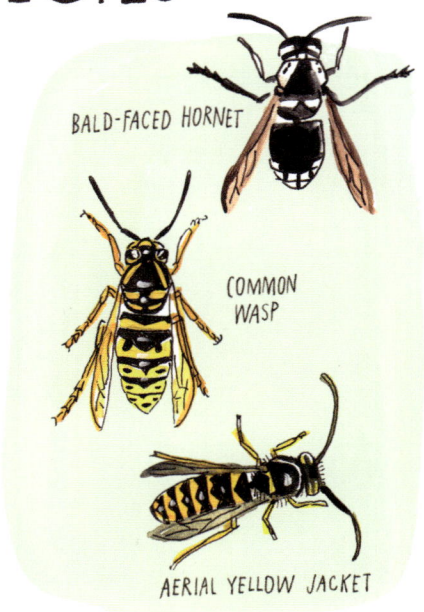

BALD-FACED HORNET

COMMON WASP

AERIAL YELLOW JACKET

SAND WASPS are solitary. They do not live in colonies. Instead, each queen will have her own nest inside a sand burrow where she lays eggs and tends to the larvae.

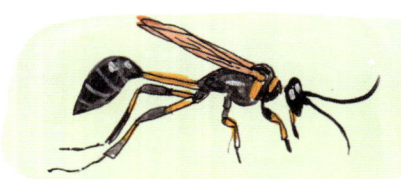

MUD DAUBER is a common name given to a number of wasp species that make their nests from molded mud. Although they are capable of stinging, mud daubers are unlikely to attack humans. Instead, their venom is typically used to paralyze and preserve their prey.

The GOLD DIGGER WASP is helpful in the garden, pollinating flowers and catching grasshoppers that can damage crops.

There are at least 19 species of YELLOW JACKET wasps, including the aerial yellow jacket, the common wasp, the eastern yellow jacket, and the bald-faced hornet.

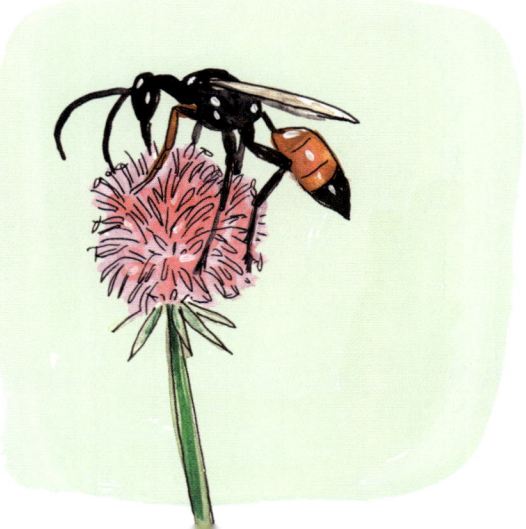

FIG WASPS and fig trees rely on one another to complete their reproductive cycles:

1. Fig wasps hatch from eggs laid in the flowers of fig trees. Those flowers are hidden inside hollow balls that develop into figs.

2. Tiny female wasps squeeze into fig flowers through miniscule openings. They die after laying their eggs, which hatch within a few days.

3. The newly hatched wasps mate within each fig. The males chew a hole through the side of the fig, but without wings, they are unable to escape and soon die.

4. The females leave the fig through the hole and fly great distances to locate other fig flowers, carrying pollen from the original fig with them to pollinate the next cycle of figs.

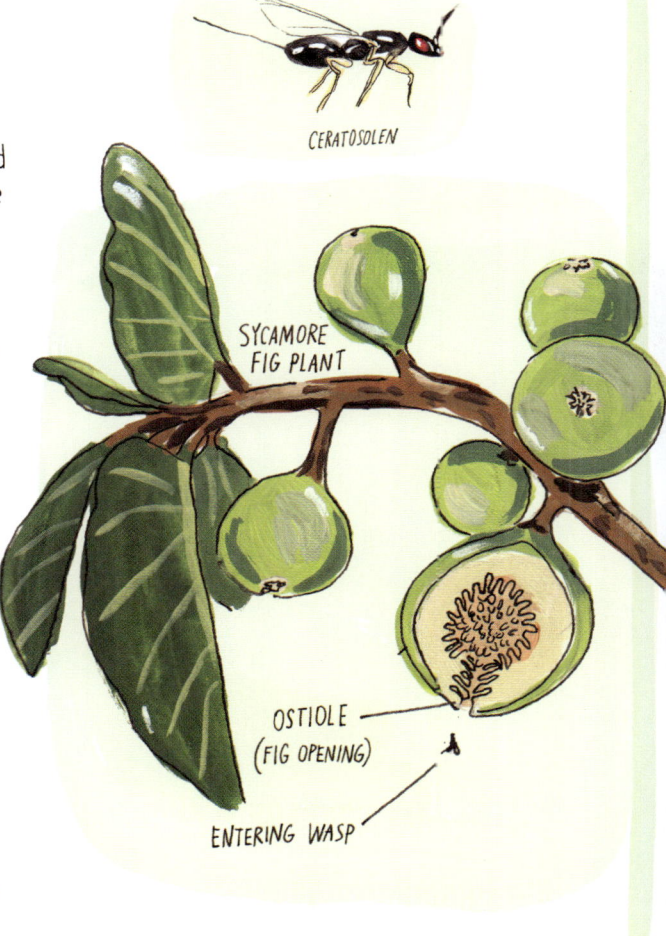

CERATOSOLEN

SYCAMORE FIG PLANT

OSTIOLE (FIG OPENING)

ENTERING WASP

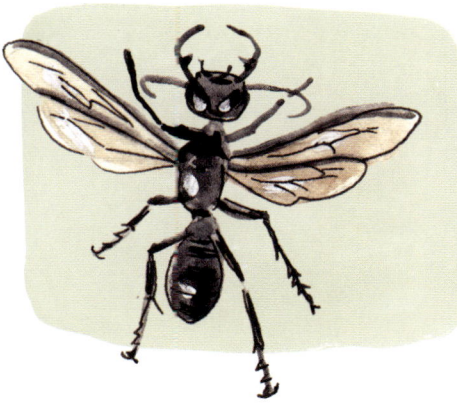

Known as the "King of Wasps," the GARUDA WASP has only been found in a particular section of the Indonesian island of Sulawesi. In fact, the species has never been observed alive or in flight. The jaws on this insect-eating wasp are so large that they wrap around the sides of the head when closed.

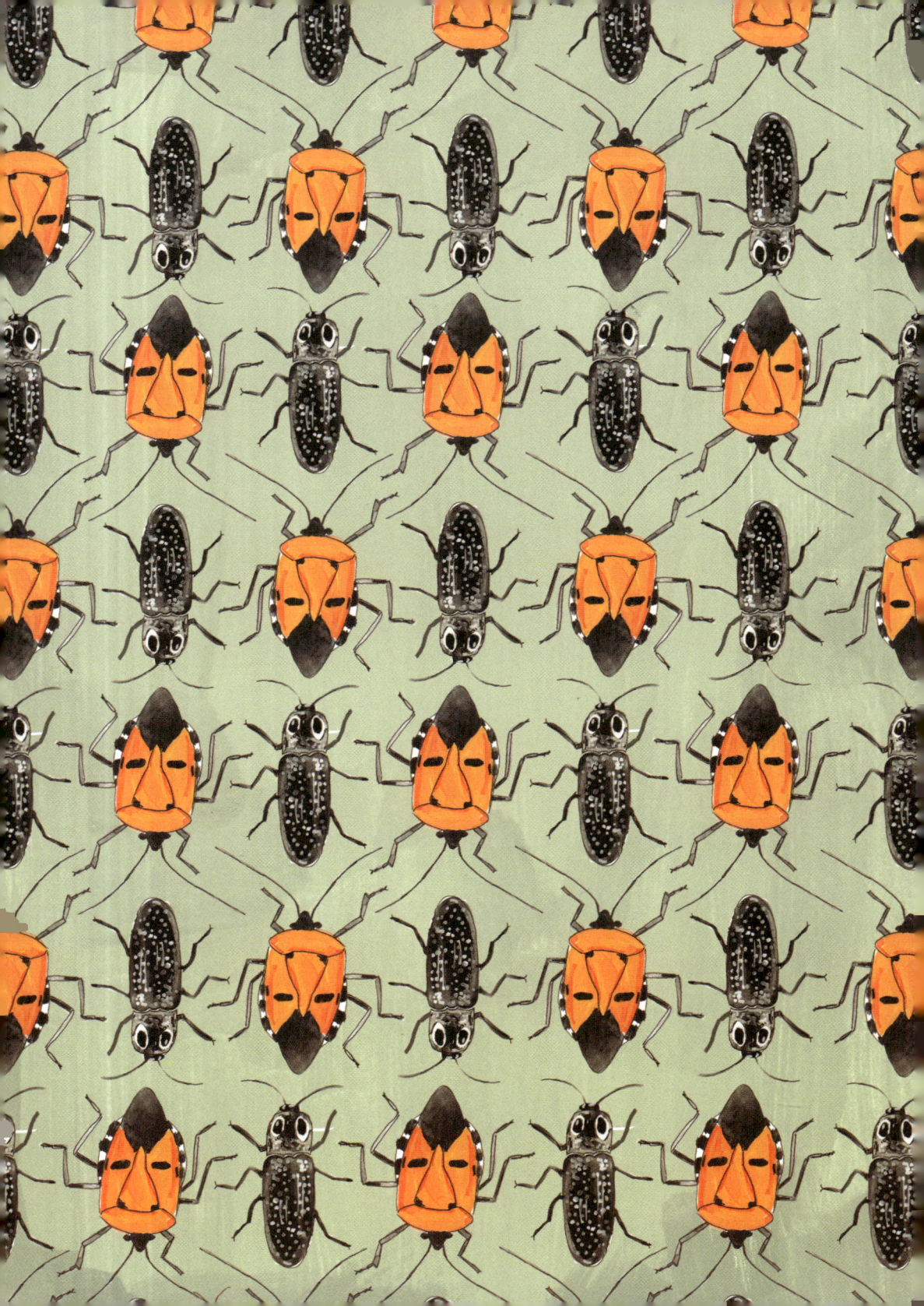

CHAPTER 4

Buzzworthy Features

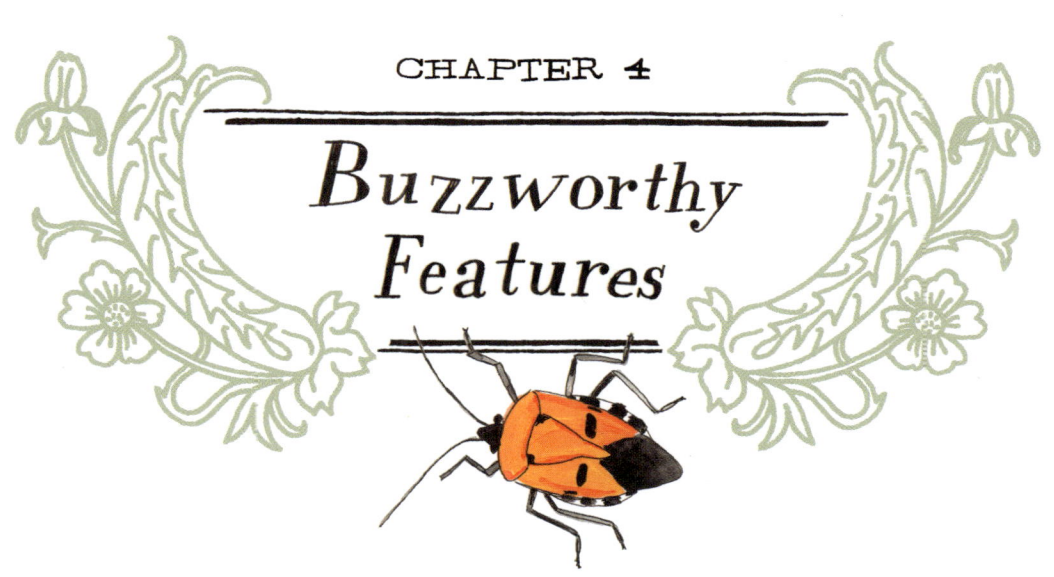

CAMOUFLAGE AND MIMICRY

Insects use various strategies to protect themselves from predators and to attract prey. Some insects are masters of disguise, using cryptic coloration, also known as camouflage, to blend in with their surroundings. Other insects use mimicry to deceive prey by resembling different objects or organisms.

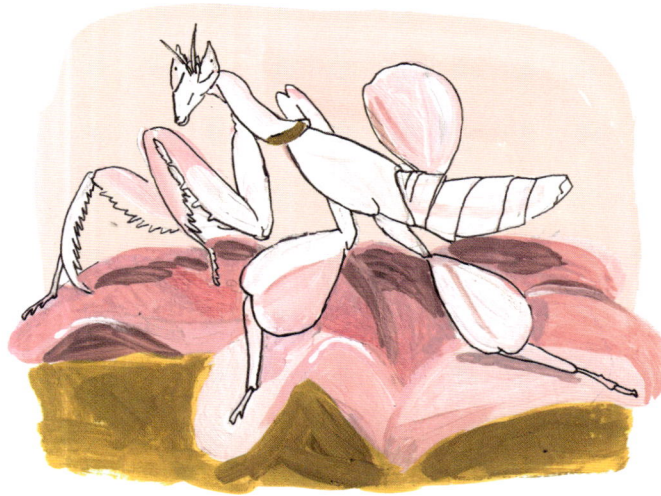

ORCHID MANTIS

Disguised as delicate flower petals, orchid mantises from Southeast Asia are in fact formidable predators capable of capturing and killing prey significantly larger than themselves.

PEPPERED MOTH

Two hundred years ago, the speckled wings of the peppered moth blended with light-colored bark. As the Industrial Revolution deposited soot over the environment, these moths evolved darker wings to match. With improved air quality, moths with lighter wings are once again common.

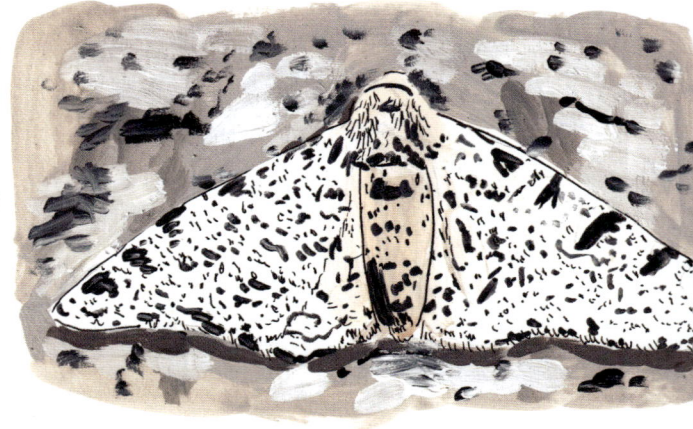

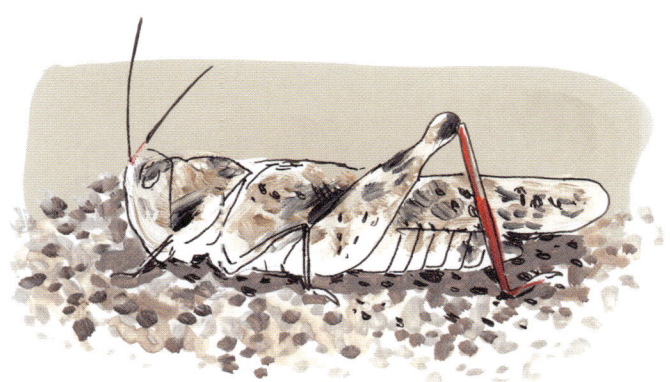

SAND GRASSHOPPER

Sand grasshoppers have coloration that helps them safely hop around in habitats with little to no grass.

WALKING LEAF

In addition to having long bodies that mimic plants, walking leaf insects will sway back and forth to emulate a gentle breeze.

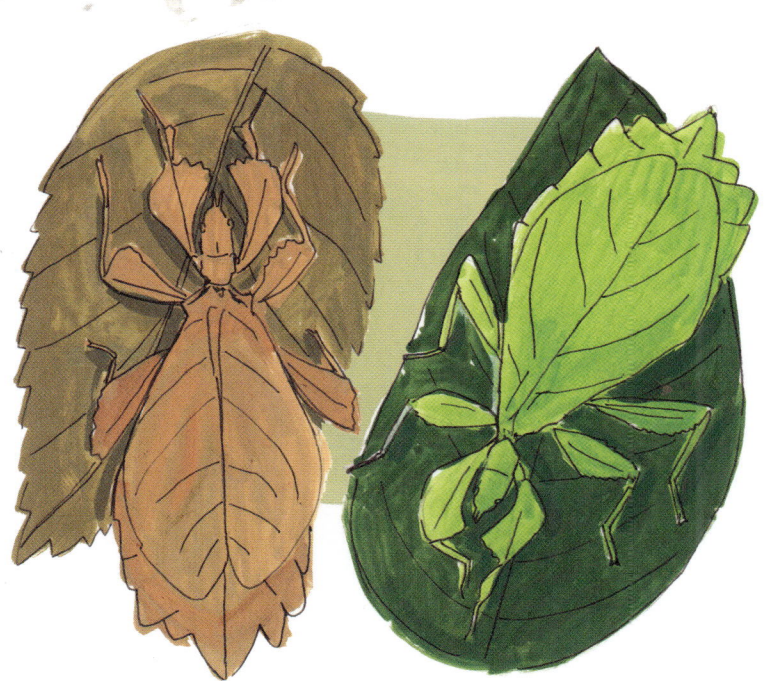

GIANT SWALLOWTAIL CATERPILLAR

Not many predators are interested in eating something that looks like bird droppings. This form of mimicry helps swallowtail caterpillars remain in the open without being eaten.

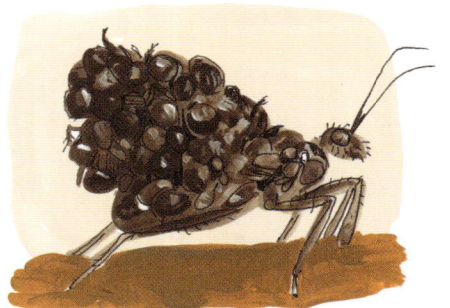

ASSASSIN BUG

Covering itself with the corpses of its victims may seem peculiar and creepy, but this form of protection makes the assassin bug significantly less vulnerable to being attacked by spiders.

THORN BUG

Thorn bugs are well disguised with their large and ornate thorax shells called pronotums. What insect predator wants to bite into a thorny branch?

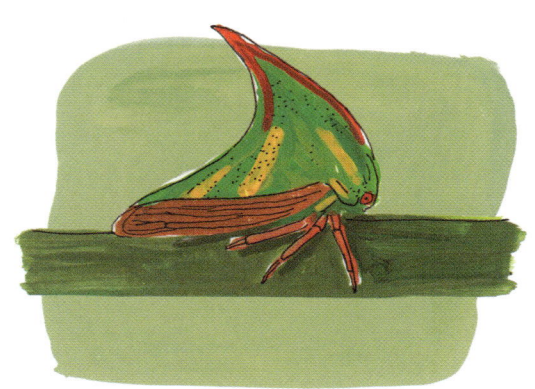

PLANTHOPPER

Despite its name, this insect prefers to stay motionless and in one place as much as possible, drawing less attention to itself. There are more than 900 species of planthoppers in North America.

KATYDID

Not only do katydids mimic leaves, but they also mimic the blemishes on leaves.

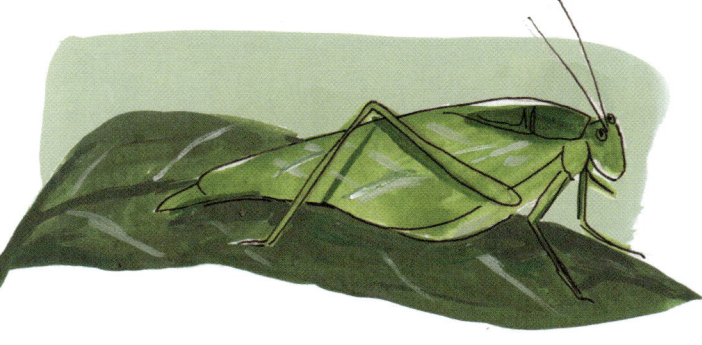

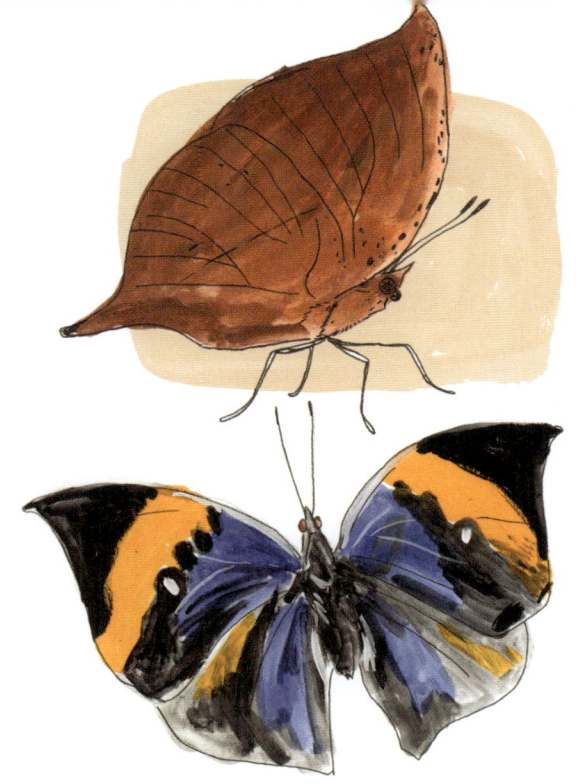

DEAD LEAF BUTTERFLY

You can see how this butterfly got its name. When its wings are closed, it can easily hide among rust-colored oak leaves. When the wings open, there's a bright, colorful surprise.

CURVE-LINED OWLET CATERPILLAR

The peculiar-looking caterpillar of the curved-line owlet moth camouflages itself as dead plant material.

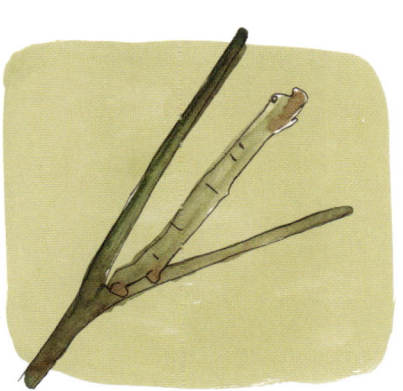

LOOPER MOTH CATERPILLAR

When threatened, these caterpillars stand erect on their front pairs of legs, increasing their resemblance to a twig.

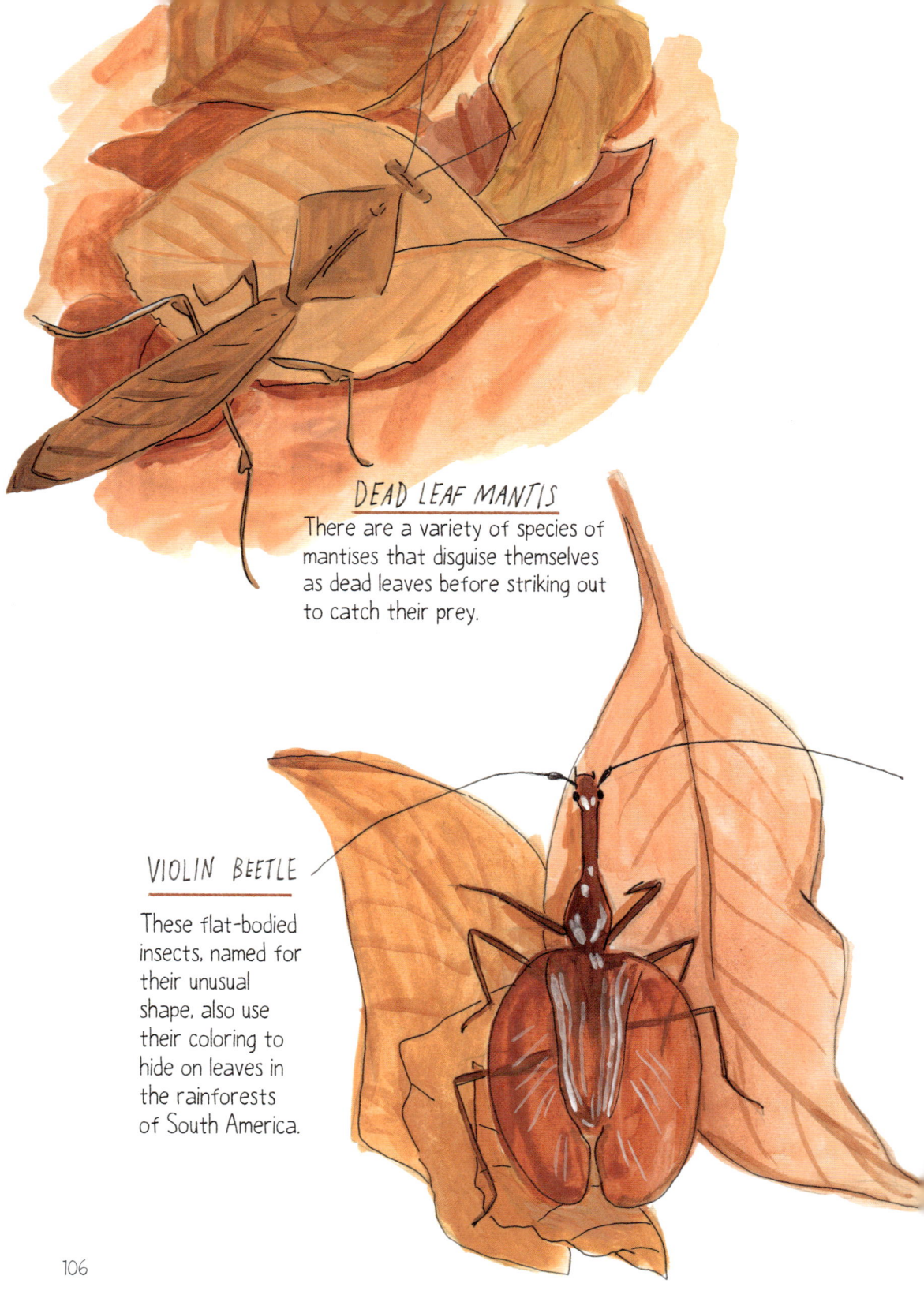

DEAD LEAF MANTIS

There are a variety of species of mantises that disguise themselves as dead leaves before striking out to catch their prey.

VIOLIN BEETLE

These flat-bodied insects, named for their unusual shape, also use their coloring to hide on leaves in the rainforests of South America.

PLANTS THAT MIMIC INSECTS

Insects mimic plants but plants also mimic insects. Some flowering plants, especially orchids, do this to attract pollinators, as a defense, or to attract prey.

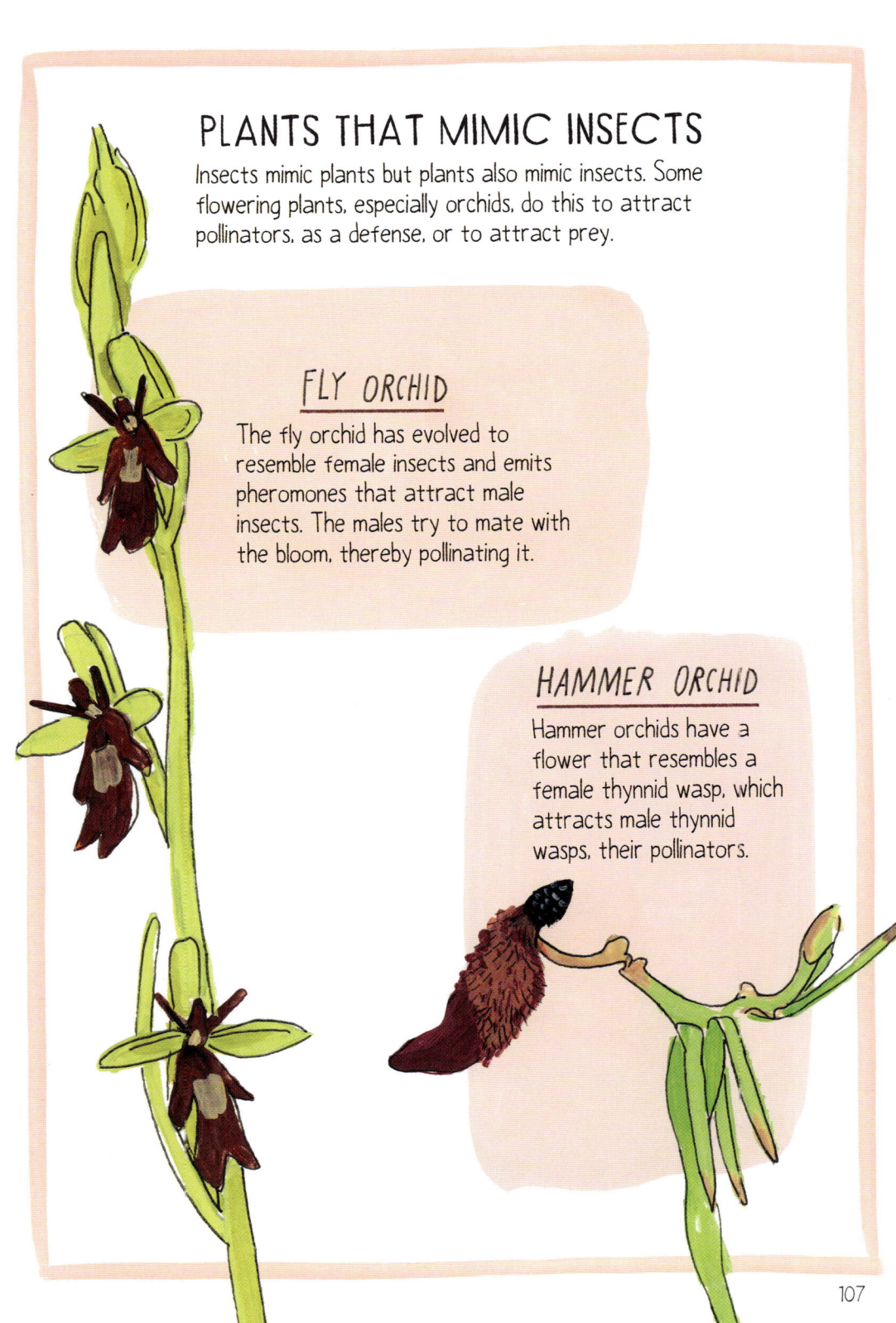

FLY ORCHID

The fly orchid has evolved to resemble female insects and emits pheromones that attract male insects. The males try to mate with the bloom, thereby pollinating it.

HAMMER ORCHID

Hammer orchids have a flower that resembles a female thynnid wasp, which attracts male thynnid wasps, their pollinators.

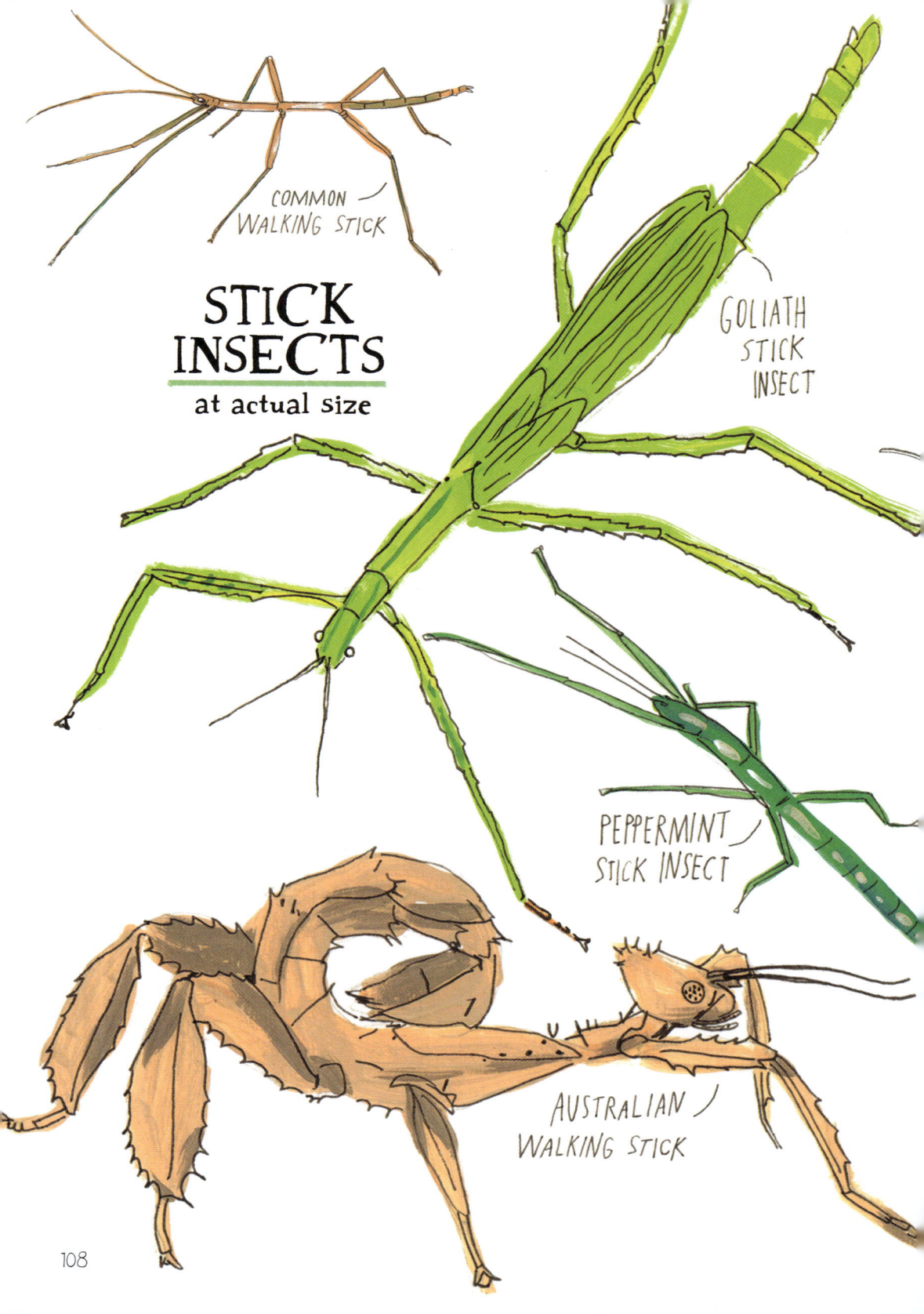

COMMON WALKING STICK

STICK INSECTS
at actual size

GOLIATH STICK INSECT

PEPPERMINT STICK INSECT

AUSTRALIAN WALKING STICK

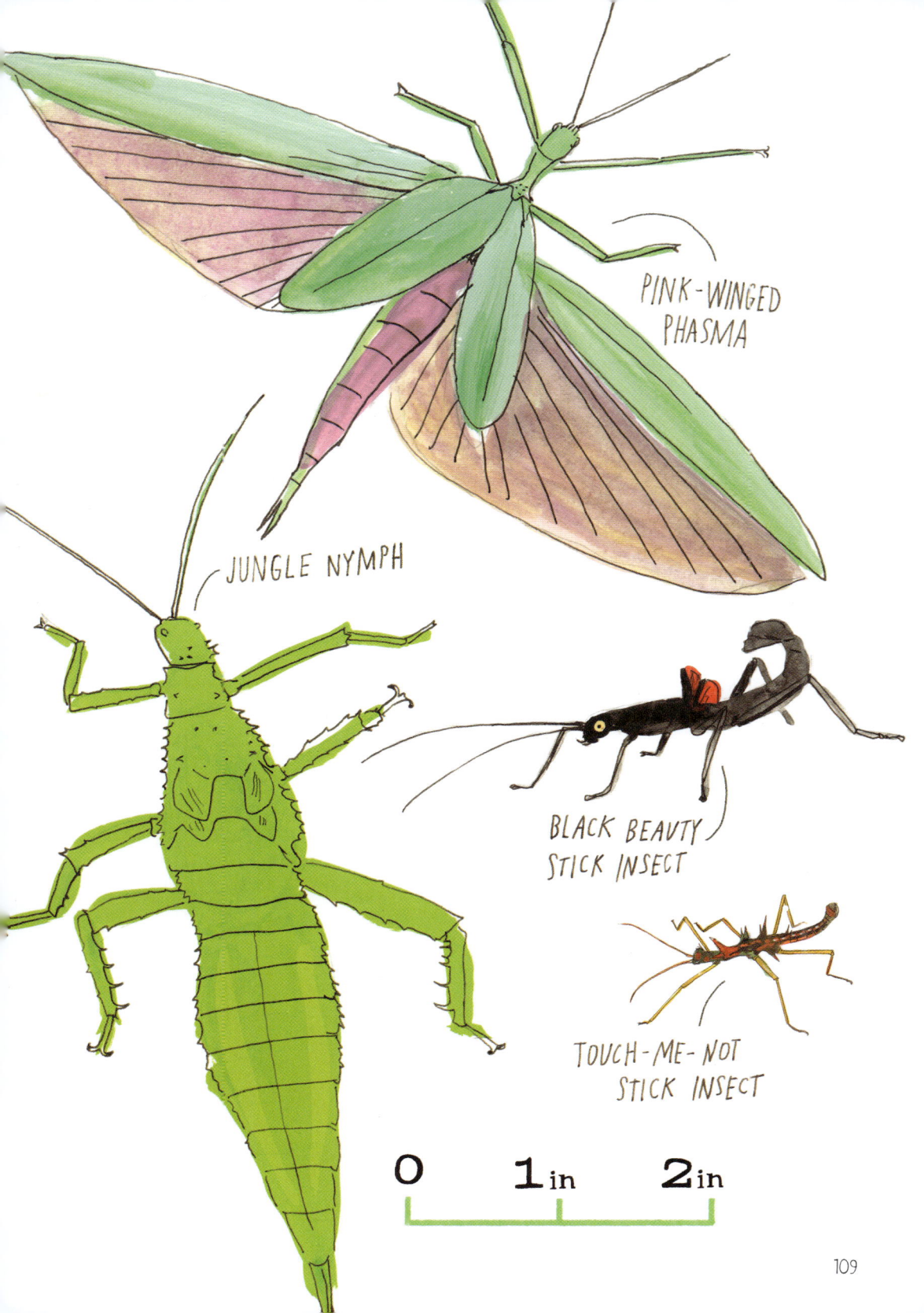

PINK-WINGED PHASMA

JUNGLE NYMPH

BLACK BEAUTY STICK INSECT

TOUCH-ME-NOT STICK INSECT

0   1in   2in

BUGS WITH FACES

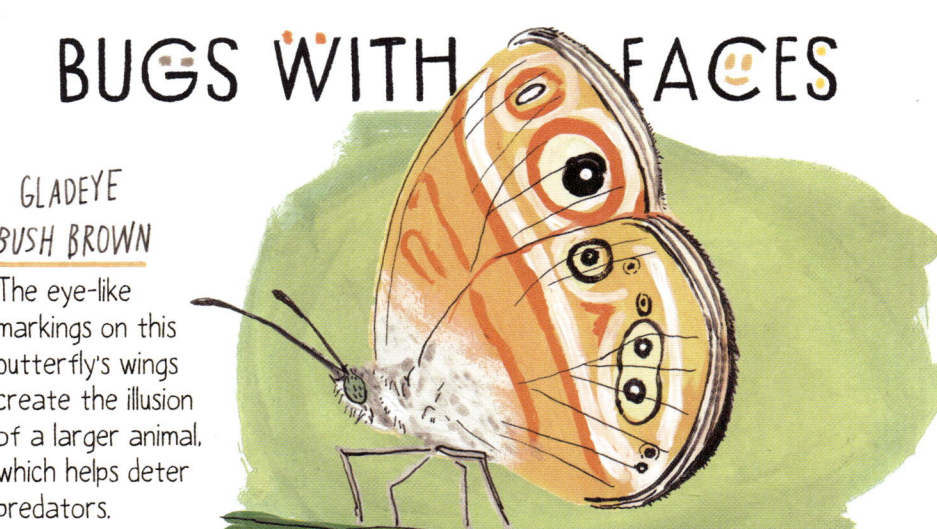

GLADEYE BUSH BROWN

The eye-like markings on this butterfly's wings create the illusion of a larger animal, which helps deter predators.

EASTERN EYED CLICK BEETLE

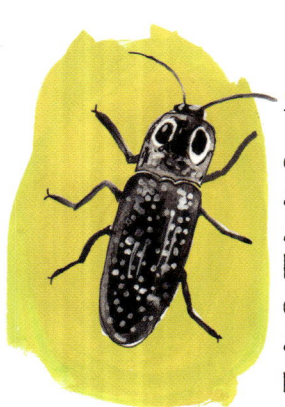

The "false eyes" on its back are a defensive adaptation that has evolved to confuse or scare away potential predators.

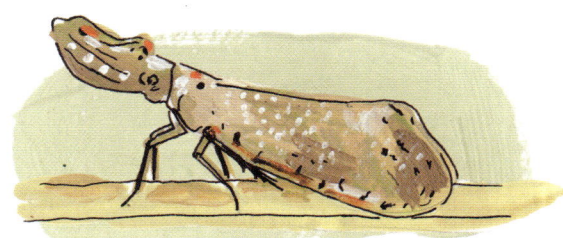

ALLIGATOR BUG

This planthopper's real head is small and almost unnoticeable compared to the fake head, which is just a built-out extension of the insect's front section.

MAN-FACED STINK BUG

Although the resemblance to a human face is completely coincidental, the markings likely ward off various predators.

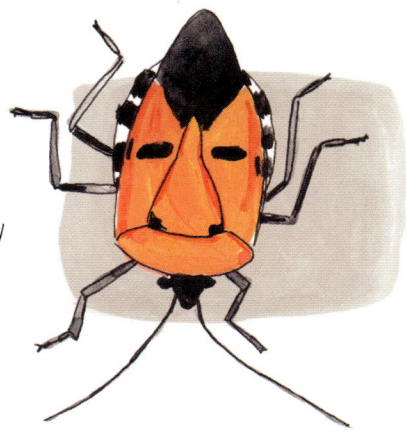

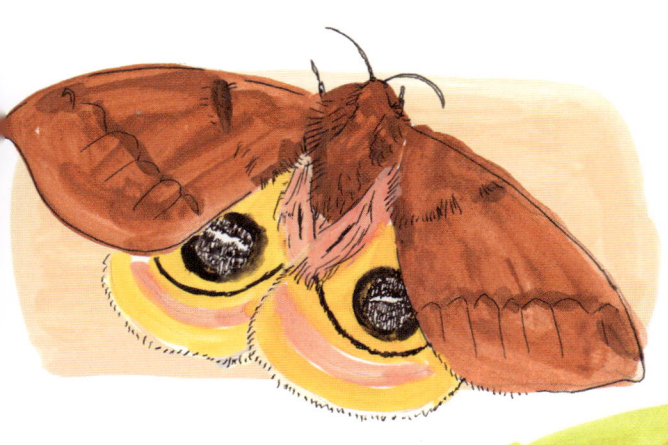

IO MOTH

The distinctive eye spots on the io moth's hindwings make it easily recognizable among other moth species.

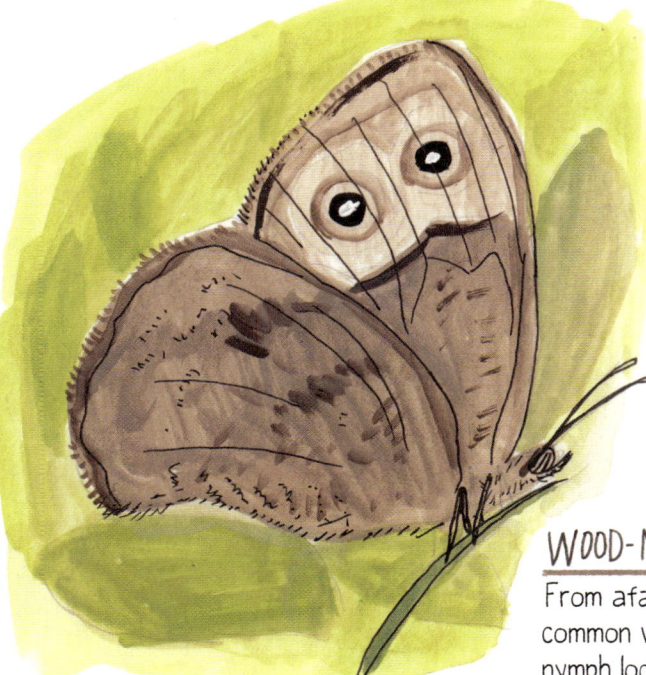

WOOD-NYMPH

From afar, the common wood-nymph looks like an owl.

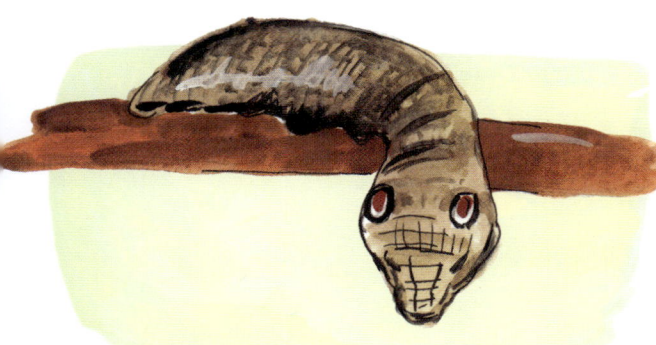

ELEPHANT HAWK MOTH CATERPILLAR

When it feels threatened, this caterpillar will wider its back to resemble a snake.

ON THE DEFENSIVE

Predators usually lose interest in dead prey; thus many insects mimic death to avoid harm, a strategy called thanatosis.

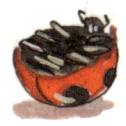

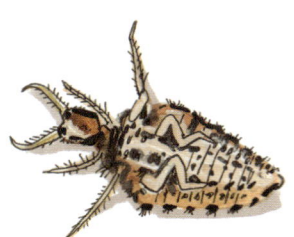

ANTLION larvae (sometimes referred to as doodlebugs) have been observed playing dead for over an hour to avoid attack.

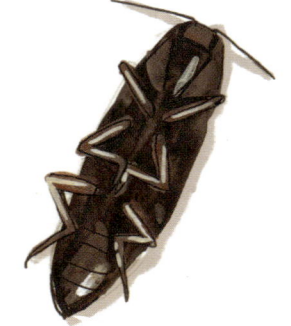

When touched, **CLICK BEETLES** dramatically fall onto their backs and pretend to be dead.

LADYBUGS protect themselves by pulling their legs up and rolling onto their backs. For added insurance, they release a small amount of blood from their legs (known as reflex bleeding), creating a bad smell that further deters predators.

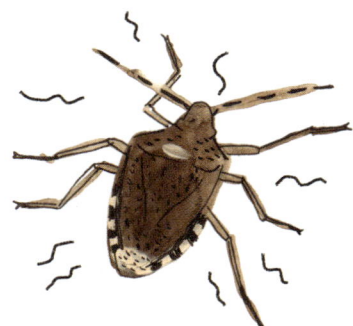

STINK BUGS release a strong and unpleasant odor that repels predators.

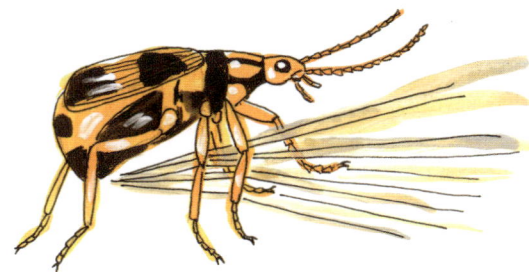

The **BOMBARDIER BEETLE** ejects a noxious chemical spray from specialized glands in its abdomen.

The Antlion Death Trap

Antlion larvae build conical pits in the sandy soil to capture ants. They wait buried under the soil with just their powerful jaws exposed. The walls of the pit are steep and the loose sand is unstable, making it easy for an ant to slip down into the powerful jaws of the hiding antlion. The larva also throws grains of sand at trapped ants, causing them to slide deeper into the funnel. Once the ant is captured, the antlion larva sucks the juices from its prey and tosses the exoskeleton from the pit to wait for its next victim.

STINGERS

Most stinging insects belong to the order Hymenoptera, which includes ants, bees, and wasps. Additionally, some caterpillars have stinging hairs or spines that can cause irritation or allergic reactions in humans and other animals.

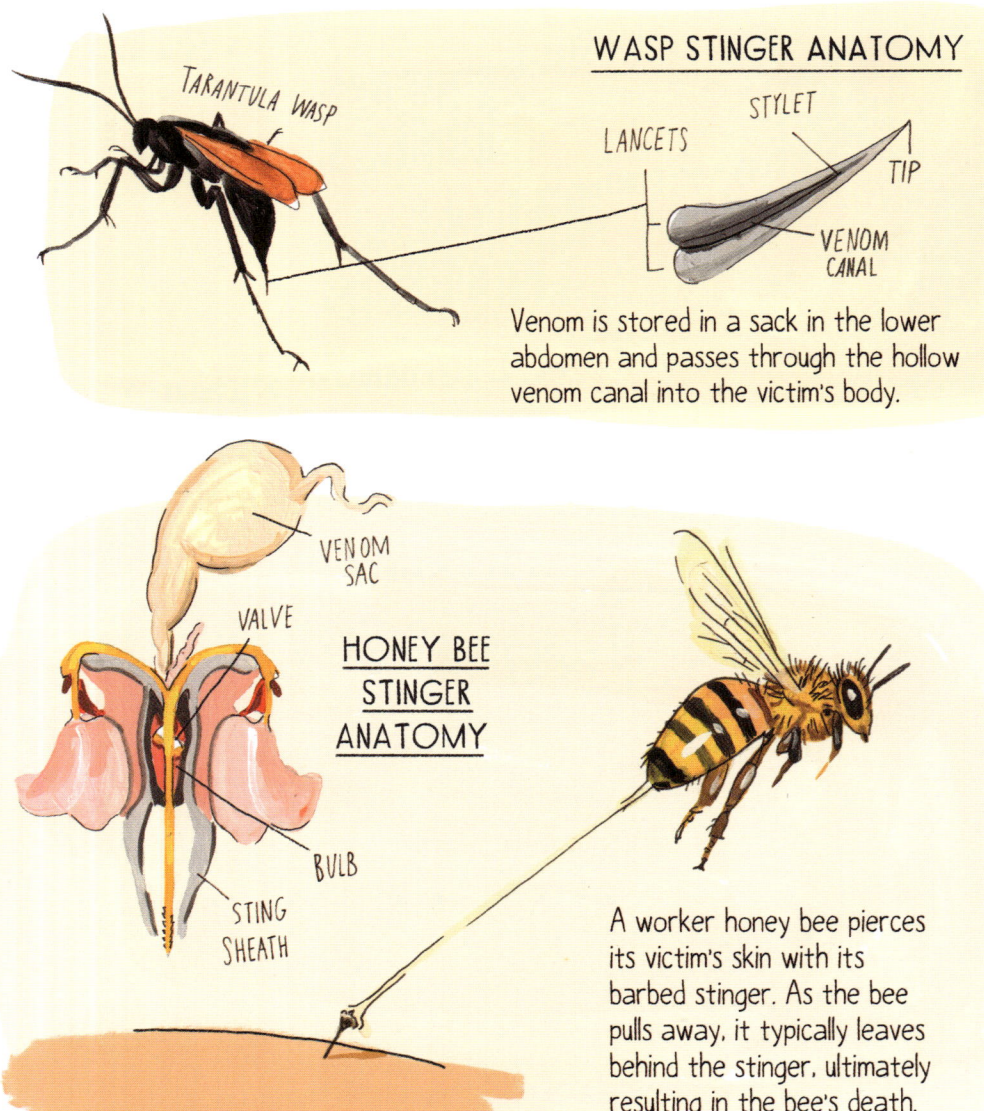

TARANTULA WASP

WASP STINGER ANATOMY

LANCETS

STYLET

TIP

VENOM CANAL

Venom is stored in a sack in the lower abdomen and passes through the hollow venom canal into the victim's body.

VENOM SAC

VALVE

HONEY BEE STINGER ANATOMY

BULB

STING SHEATH

A worker honey bee pierces its victim's skin with its barbed stinger. As the bee pulls away, it typically leaves behind the stinger, ultimately resulting in the bee's death.

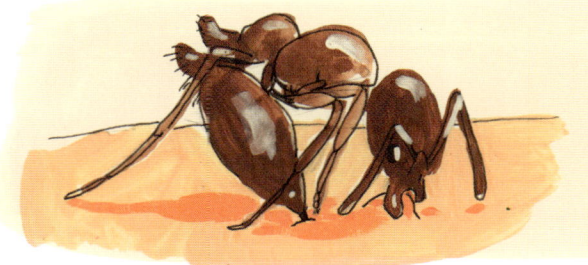

The fire ant bites first to anchor itself before thrusting its stinger into the skin.

Some species of caterpillar have spiny hairs that are connected to poisonous glands.

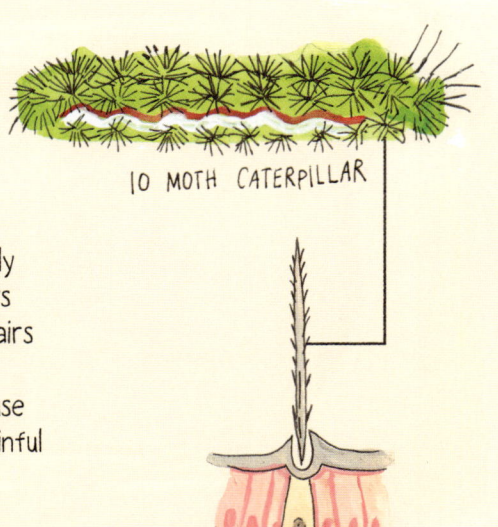

IO MOTH CATERPILLAR

These urticating hairs can easily break off from the caterpillar's body if brushed against. The hairs are equipped with tiny barbs and/or chemicals that can cause irritation, itching, or even a painful rash upon contact.

Schmidt's Sting Pain Index, created by entomologist Justin O. Schmidt, rates the pain of insect stings on a scale from 1 to 4 (4 being the most painful). Here are some insect sting ratings.

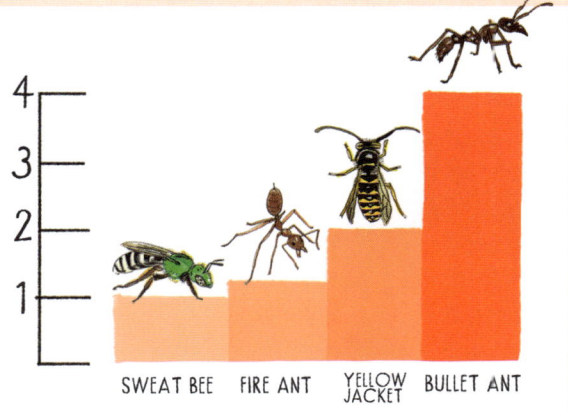

4
3
2
1

SWEAT BEE   FIRE ANT   YELLOW JACKET   BULLET ANT

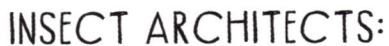

INSECT ARCHITECTS:

The Shelter Builders

EASTERN TENT CATERPILLARS

To protect themselves against chilly early spring temperatures, eastern tent caterpillars weave large 3D webs using several branches as anchor points.

LEAFROLLERS

Leafrollers, the larvae of certain tortricid moths, frequently roll themselves into leaves to feed and transform into pupae.

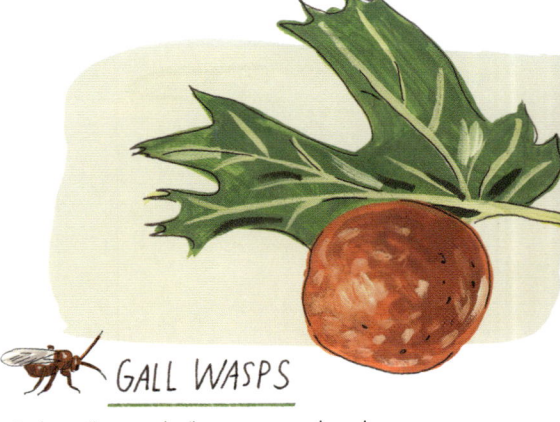

GALL WASPS

Oak galls are bulbous growths that form when gall wasps lay their eggs on oak trees. The chemicals released by the larvae cause the tree to produce the gall, which provides both protection and nourishment for the developing larvae.

WEAVER ANTS

Weaver ants work cooperatively to pull together the edges of leaves to create large nests. Worker ants hold larvae to the joined leaf edges, tapping their heads to cause them to excrete silk to create a seal.

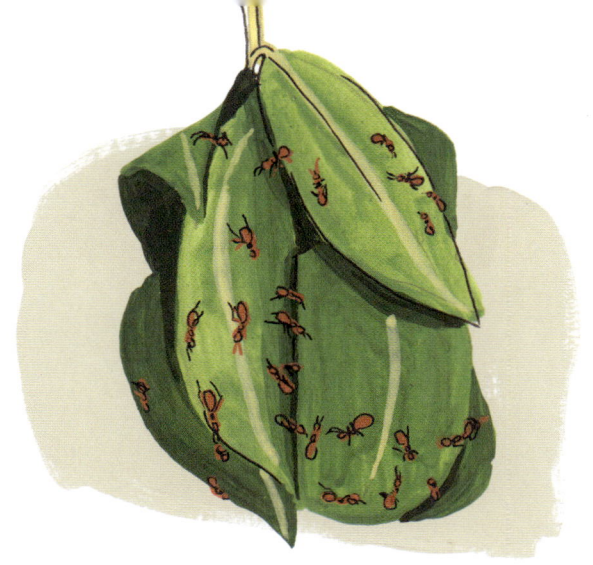

HARVESTER ANTS

Harvester ants in India build nests with an outside structure of concentric walls. It is believed that this construction helps keep rain out of the nest during monsoon season.

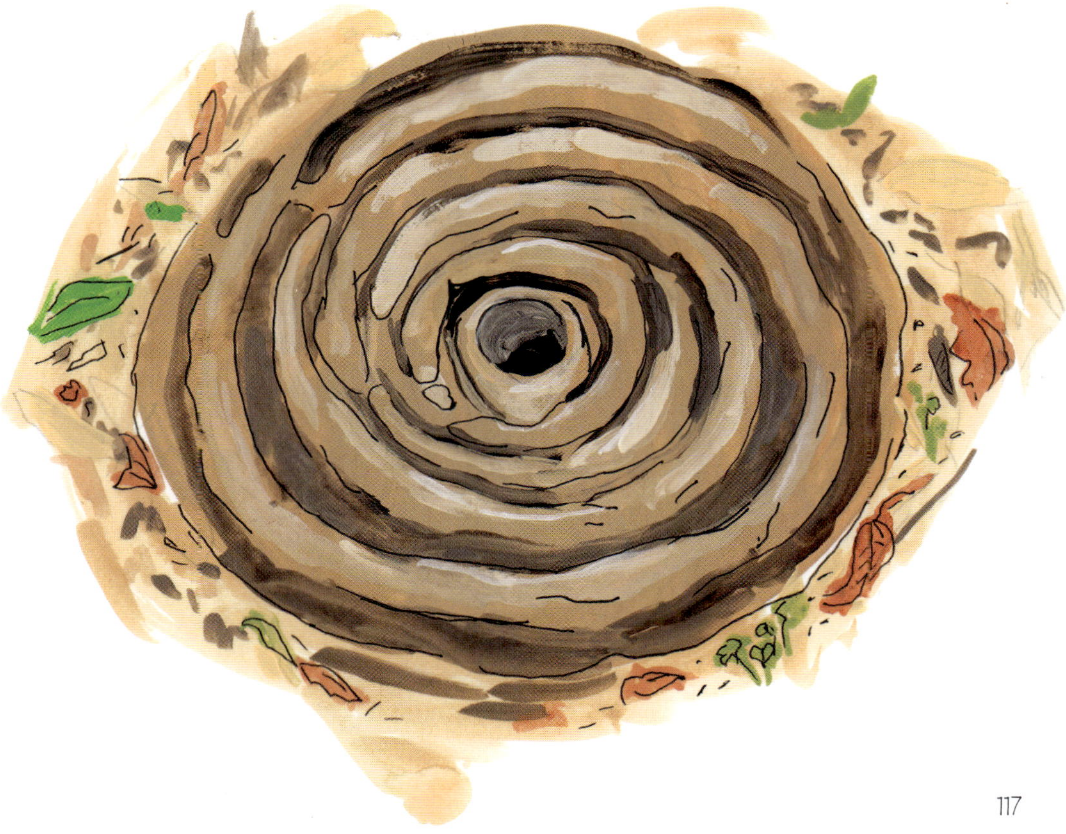

The Channel Builders

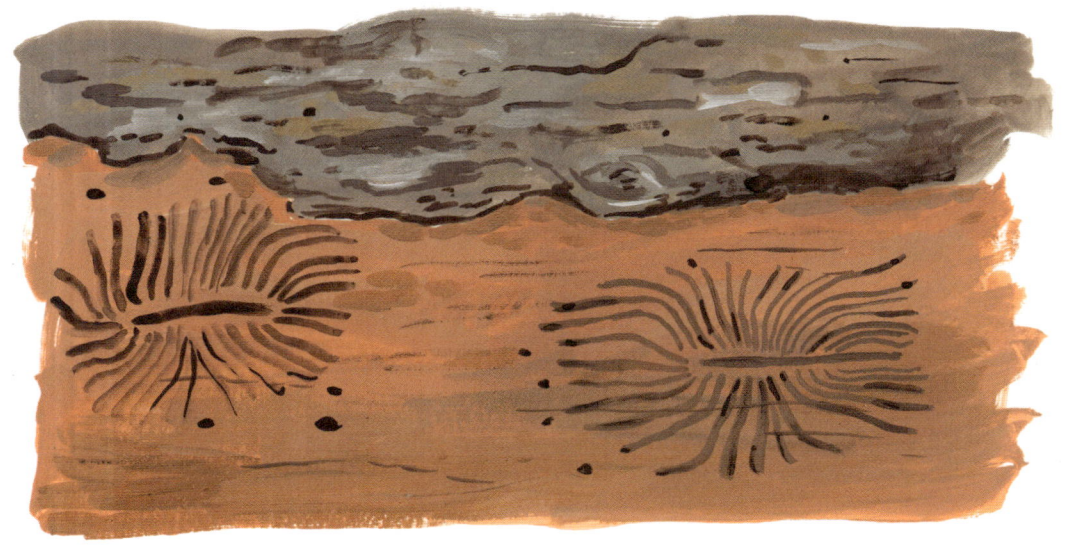

BARK BEETLES

Bark beetles have powerful mandibles that allow them to chew through tree bark to lay their eggs. Once the larvae hatch, they feed on the inner bark, creating winding, feathery tunnels. While bark beetles typically target dead or dying trees, a surge in their population—possibly linked to climate change—has led them to infest healthy trees as well, causing those trees to die.

LEAF MINERS

Leaf-mining insects create intricate tunnels and patterns within the leaves they inhabit, feeding on the inner tissues and often leaving visible trails.

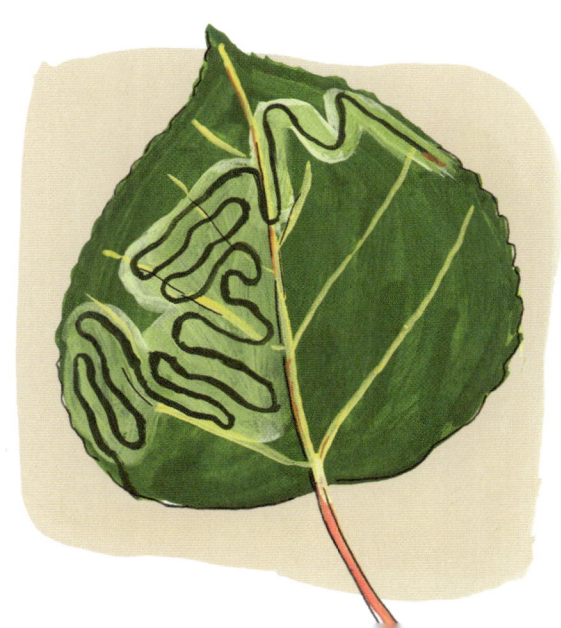

The Armor Builders

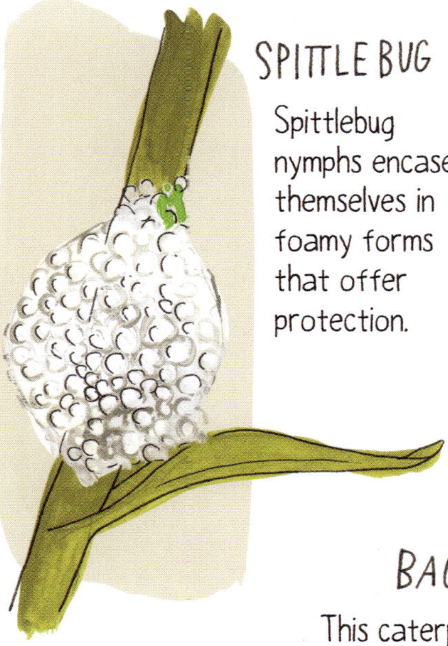

SPITTLE BUG

Spittlebug nymphs encase themselves in foamy forms that offer protection.

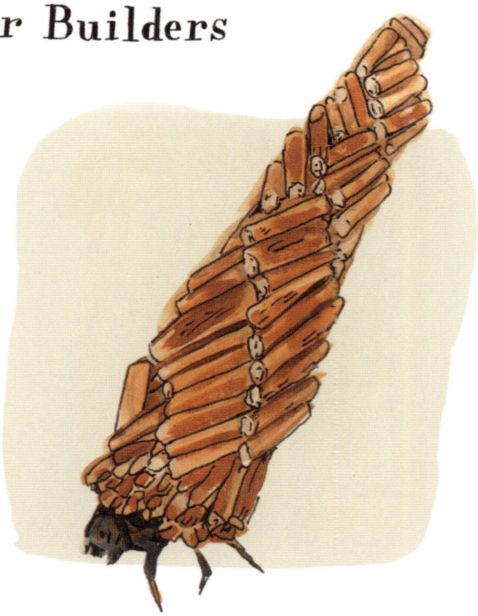

BAGWORM MOTH CATERPILLAR

This caterpillar constructs a mobile home around itself made of twigs and leaves. It carries the structure, which resembles a tiny log cabin, with it as it grows.

CASEBEARER & CADDISFLY LARVAE

These creatures safeguard themselves by creating a casing of sand, pebbles, plant bits, and other debris gathered from their surroundings.

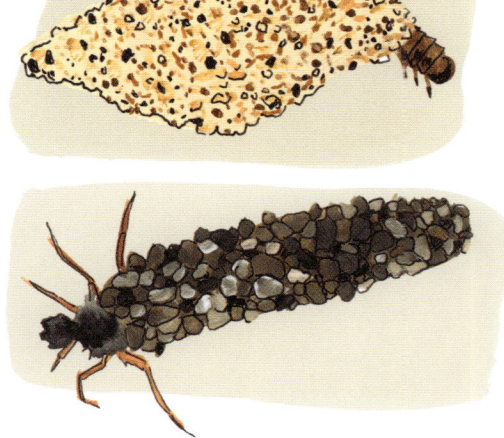

BIG JUMPERS

Grasshoppers are characterized by their powerful hind legs, which give them the ability to catapult through the air.

Grasshoppers can jump 10-20 times their body length, which would be the equivalent of a human jumping over a football field.

In addition to jumping, grasshoppers are also strong fliers and can escape predators by taking wing.

ANATOMY OF A GRASSHOPPER

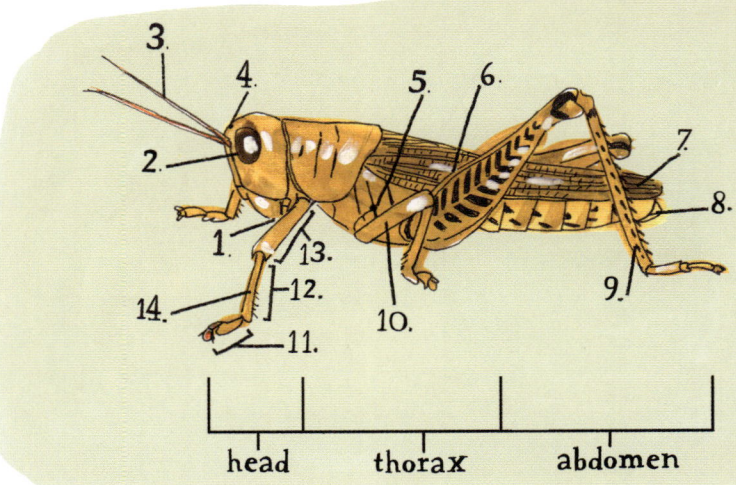

head   thorax   abdomen

1. mandible
2. compound eye
3. antenna
4. simple eye
5. spiracle
6. tympanum
7. wings

8. ovipositor
9. hindleg
10. middle leg
11. tarsus
12. tibia
13. femur
14. foreleg

GRASSHOPPER SPECIES

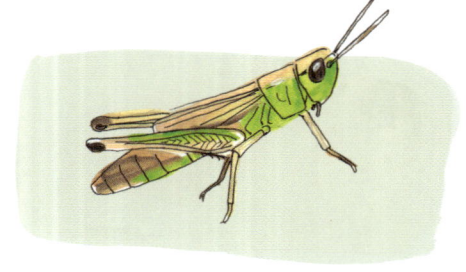

MEADOW

These insects rarely venture far from their grassy home. While the males can jump about 10 feet in a single hop, both sexes have short wings rendering them incapable of flight.

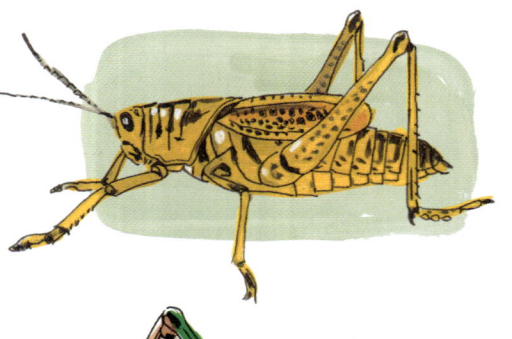

LUBBER

This large grasshopper can be identified by its distinct coloration, which ranges from black to yellow and orange depending on its age.

MIGRATORY LOCUST

Locusts are a type of grasshopper. When they enter a swarming phase, they develop more powerful hindlegs and longer wings that enable them to fly long distances.

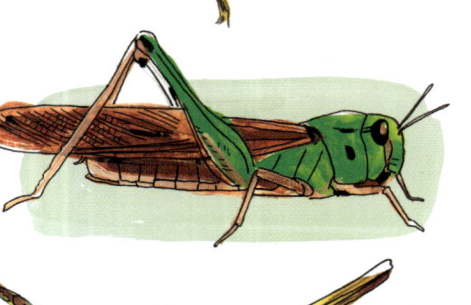

CHINESE

Commonly found in the grasslands and tropical forests of China, Japan, and Southeast Asia, this locust has a distinctly long triangular head.

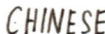

RED LOCUST

This brightly colored grasshopper is found in sub-Saharan Africa. The female can grow to be more than 3 inches long.

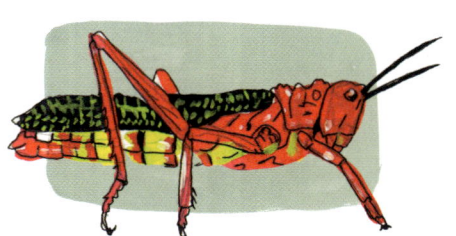

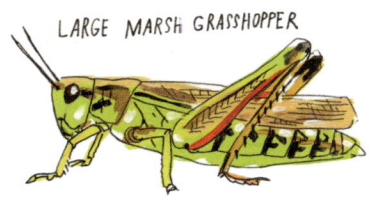

LARGE MARSH GRASSHOPPER

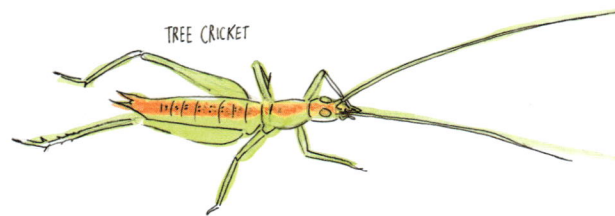

TREE CRICKET

GRASSHOPPER VS. CRICKET

SHORT ANTENNAE	LONG ANTENNAE
SHORT, THICK WINGS	LONG, THIN WINGS
RUBS LEGS FOR BUZZING SOUNDS	RUBS WINGS TO CHIRP
DIURNAL	NOCTURNAL
FOUND IN DRY FIELDS AND MEADOWS	FOUND IN MOIST DARK PLACES

CRICKETS AS THERMOMETERS

Amazingly, the frequency of a cricket's chirping changes with the temperature. To estimate the temperature in degrees Fahrenheit, count the number of chirps in 15 seconds and add 37. The resulting number is an approximate measure of the outside temperature!

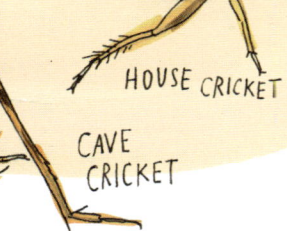

HOUSE CRICKET

FIELD CRICKET

CAVE CRICKET

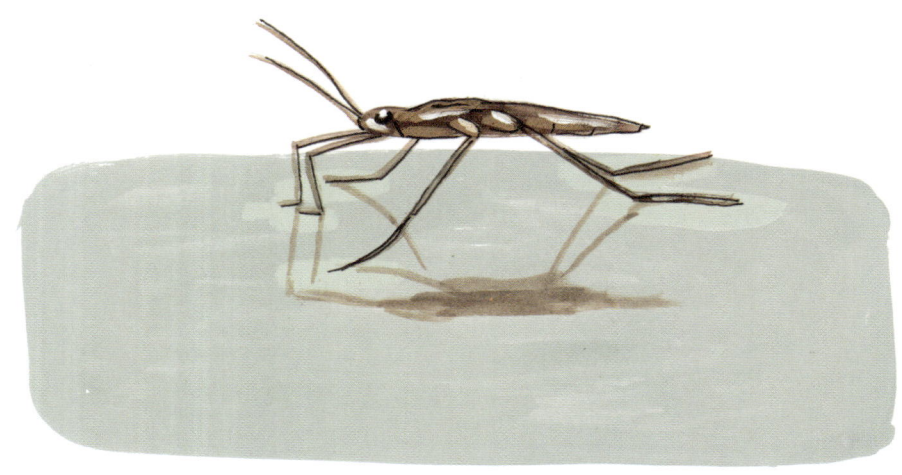

WALKING ON WATER

Several species of insects, particularly in the family Gerridae, can walk or move across the surface of water due to their unique adaptations and abilities. They are often called water striders, pond skaters, or puddle flies.

With their long, slender legs, water striders can distribute their weight over a large surface area. Their legs are covered in water-repellent hairs that prevent the insect from breaking the surface tension of the water.

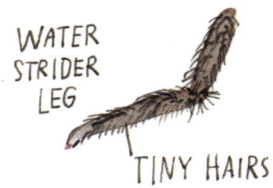

WATER STRIDER LEG

TINY HAIRS

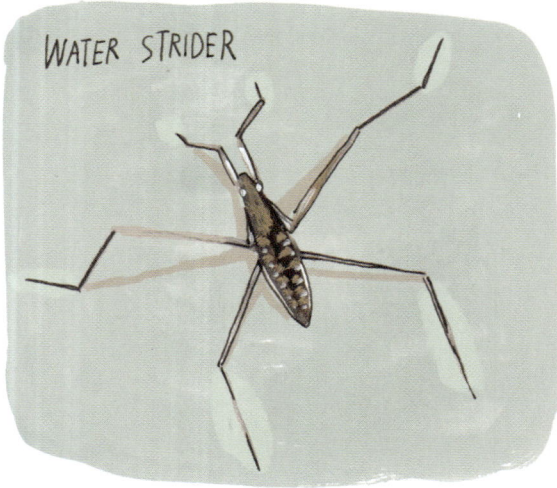

WATER STRIDER

Water striders use their middle legs to row or paddle across the water, creating ripples that propel them forward. Their shorter front legs are for capturing prey and their hindlegs are used for steering and balance.

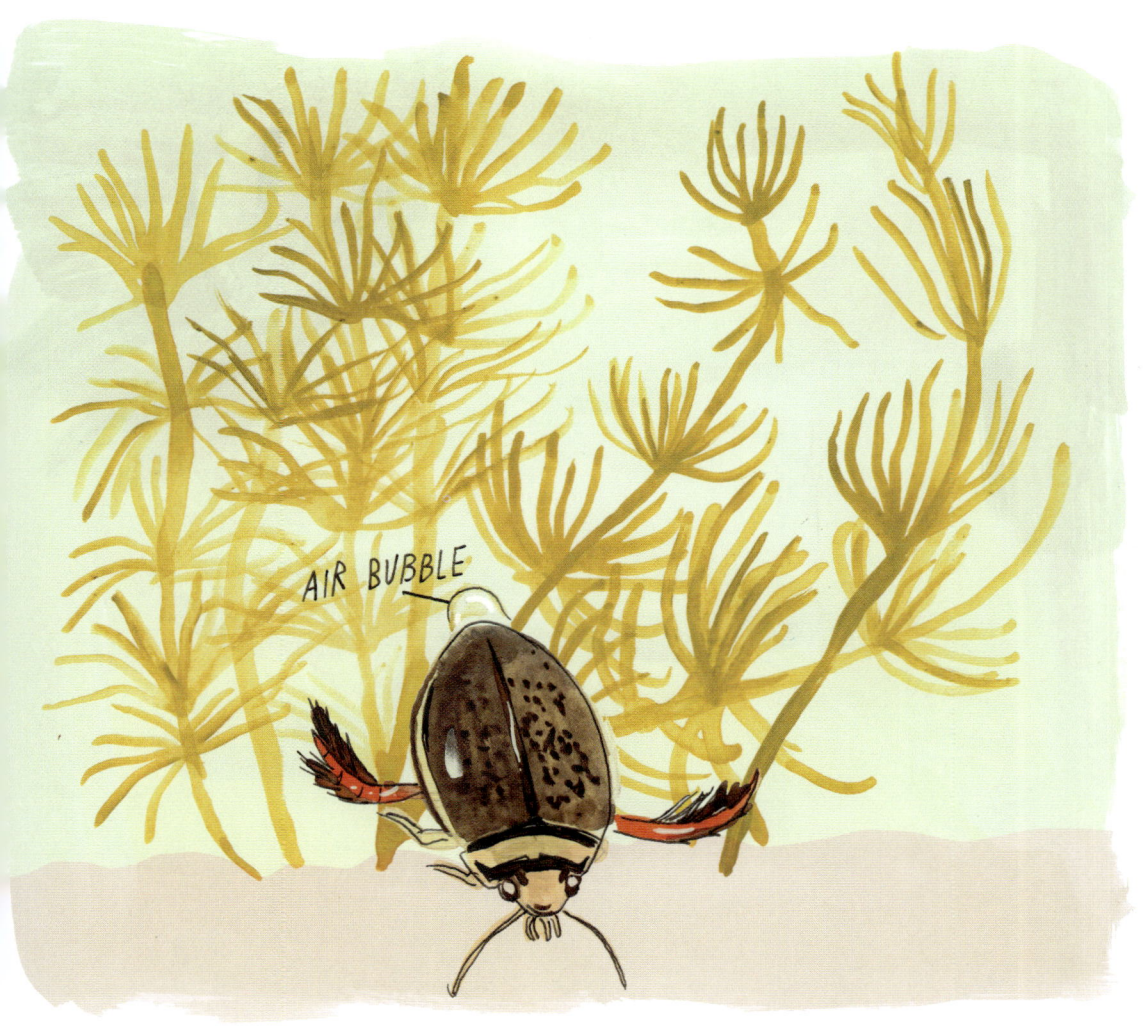

AIR BUBBLE

DIVING DEEP

Predaceous diving beetles have streamlined bodies and strong, fringed legs for swimming. They can stay underwater for up to 30 minutes, carrying a supply of air in the form of a bubble under their elytra, or wings.

These fierce hunters, no more than 1½ inches, will attack tadpoles, small fish, and even frogs, grasping them with sharp pincers that inject digestive enzymes into their prey.

Larvae, called "water tigers," are also strong predators.

PLANT + INSECT COLLABORATORS

Some plants and insects have developed mutually beneficial relationships for protection, pollination, and seed dispersal.

Milkweed + Monarchs

Monarch caterpillars rely on milkweed leaves as their sole food source. The leaves contain toxic compounds that make both the caterpillars and the butterflies distasteful to predators. In turn, monarch butterflies pollinate milkweed plants as they feed on the nectar and fly from plant to plant.

Yucca Plant + Yucca Moth

The yucca moth is the only insect that can pollinate this plant's flowers, making it possible for the plant to produce seeds. In return, the yucca plant provides a safe place from predators or bad weather for the moth to lay its eggs. When the eggs hatch, the larvae feed on the yucca seeds.

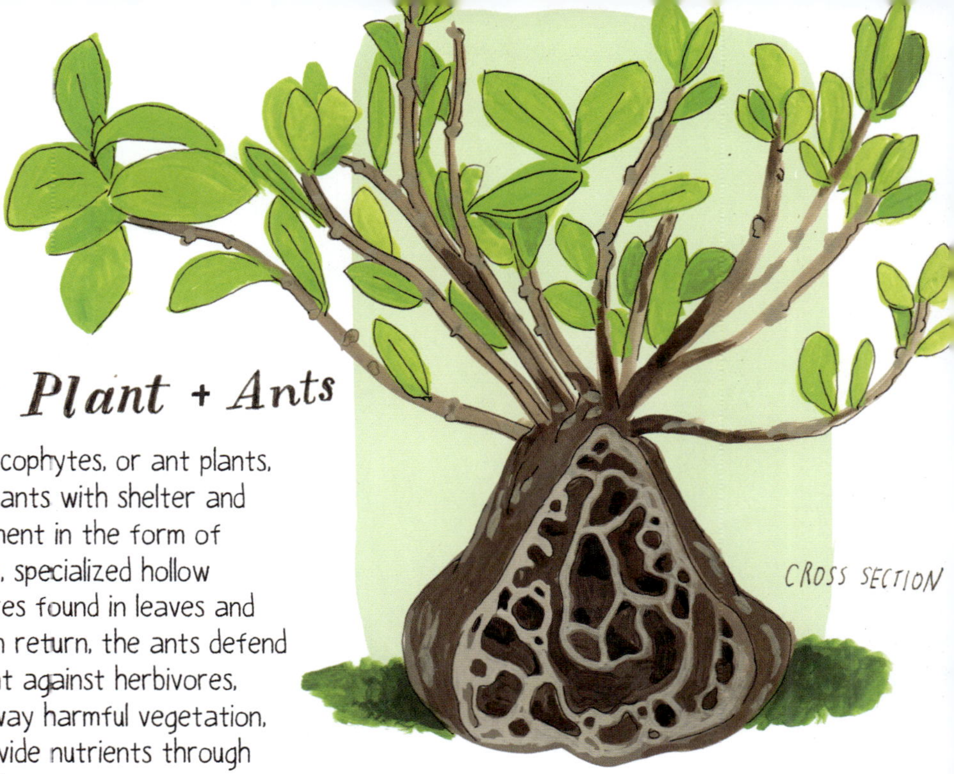

Ant Plant + Ants

Myremecophytes, or ant plants, provide ants with shelter and nourishment in the form of domatia, specialized hollow structures found in leaves and stems. In return, the ants defend the plant against herbivores, clear away harmful vegetation, and provide nutrients through debris piles.

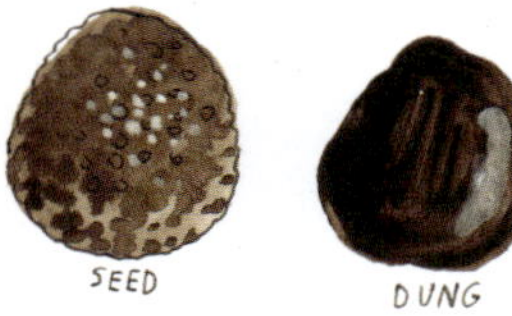

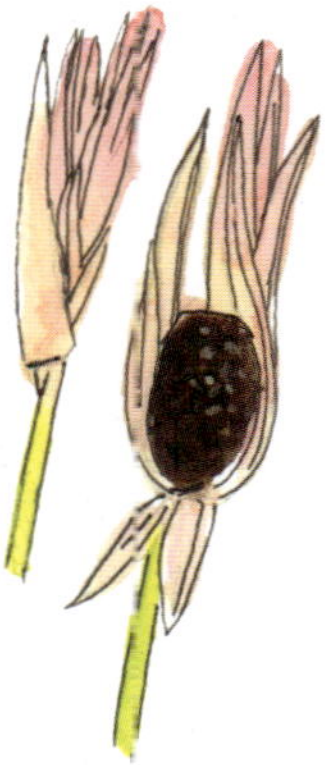

CROSS SECTION

SEED

DUNG

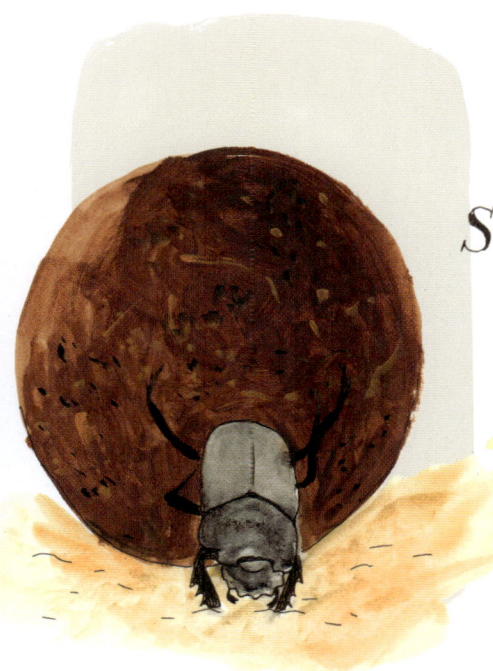

Silver Arrowreed + Dung Beetle

Some species of dung beetles roll manure into balls and bury it, which recycles nutrients into the soil and helps plants grow. In South Africa, the silver arrowreed plant has large, pungent seeds that resemble dung balls, tricking the beetles into burying them. This strategy helps disperse the plant into new areas.

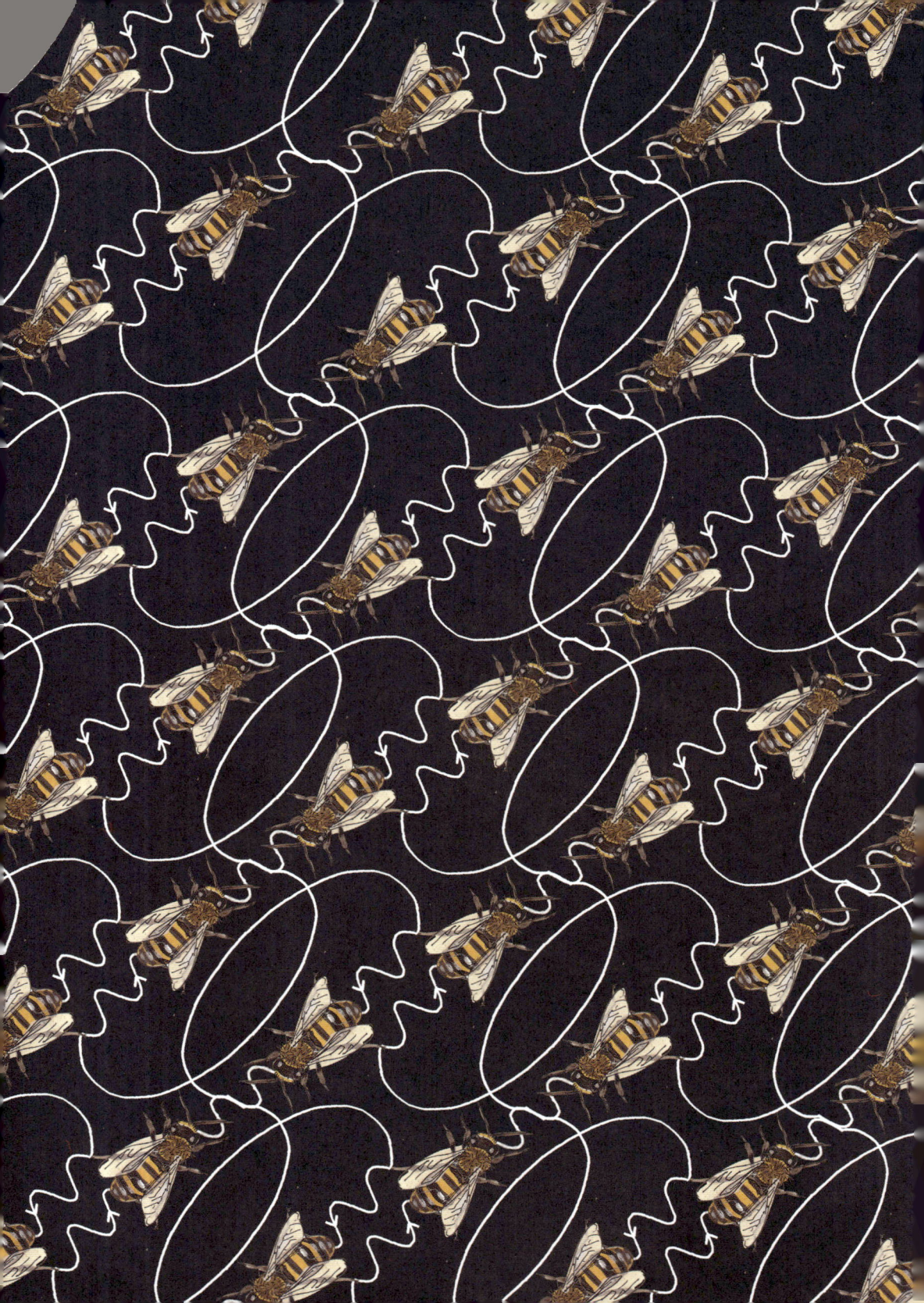

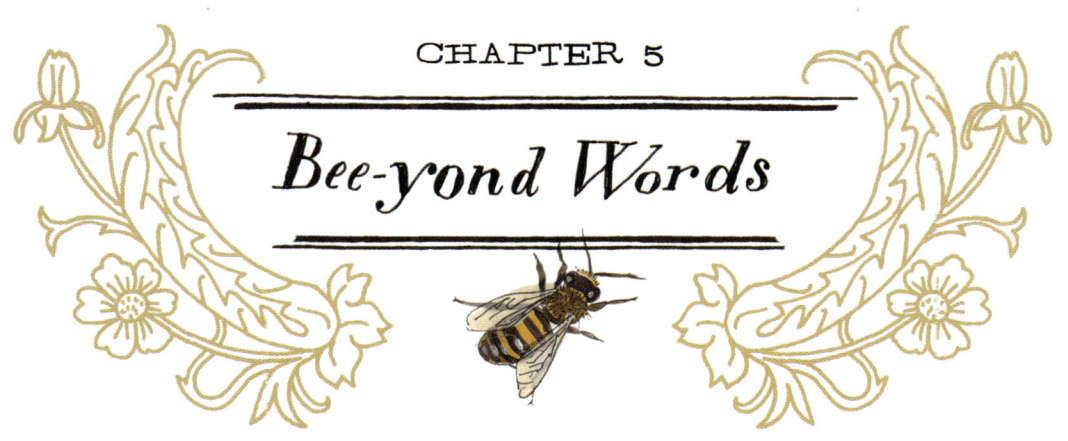

CHAPTER 5

Bee-yond Words

WAGGLE, WAGGLE

Honey bees communicate the location of food sources through a complex series of movements known as the "waggle dance." The dance conveys information about the direction and distance of the food in relation to the sun's position.

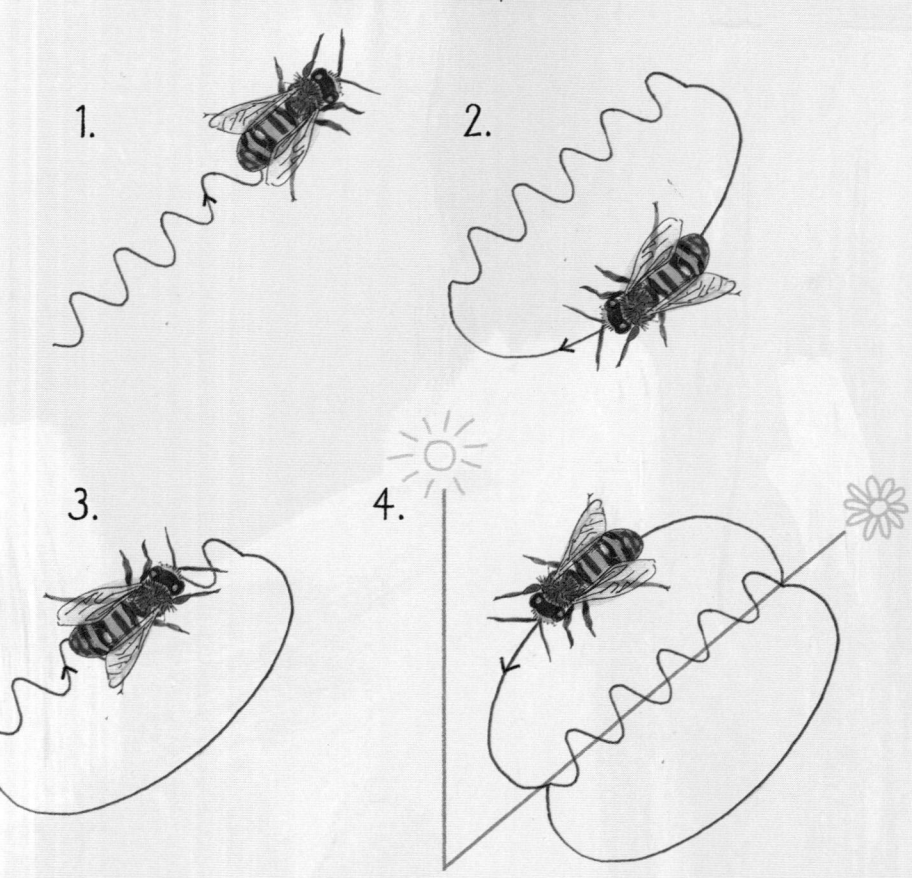

Inside the hive, the bee dances forward while waggling from side to side, then circles back to repeat the motion. The duration of the waggle indicates the approximate distance to the food source, while the angle of the dance relative to the top of the hive matches the angle of the food source in relation to the sun.

STRIKE A POSE

Insects communicate in various ways, some through their behaviors and the release of chemical odors, while others produce sounds, emit light, or rely on coloration. These communication methods allow insects to interact, share information, and navigate their surroundings with incredible precision.

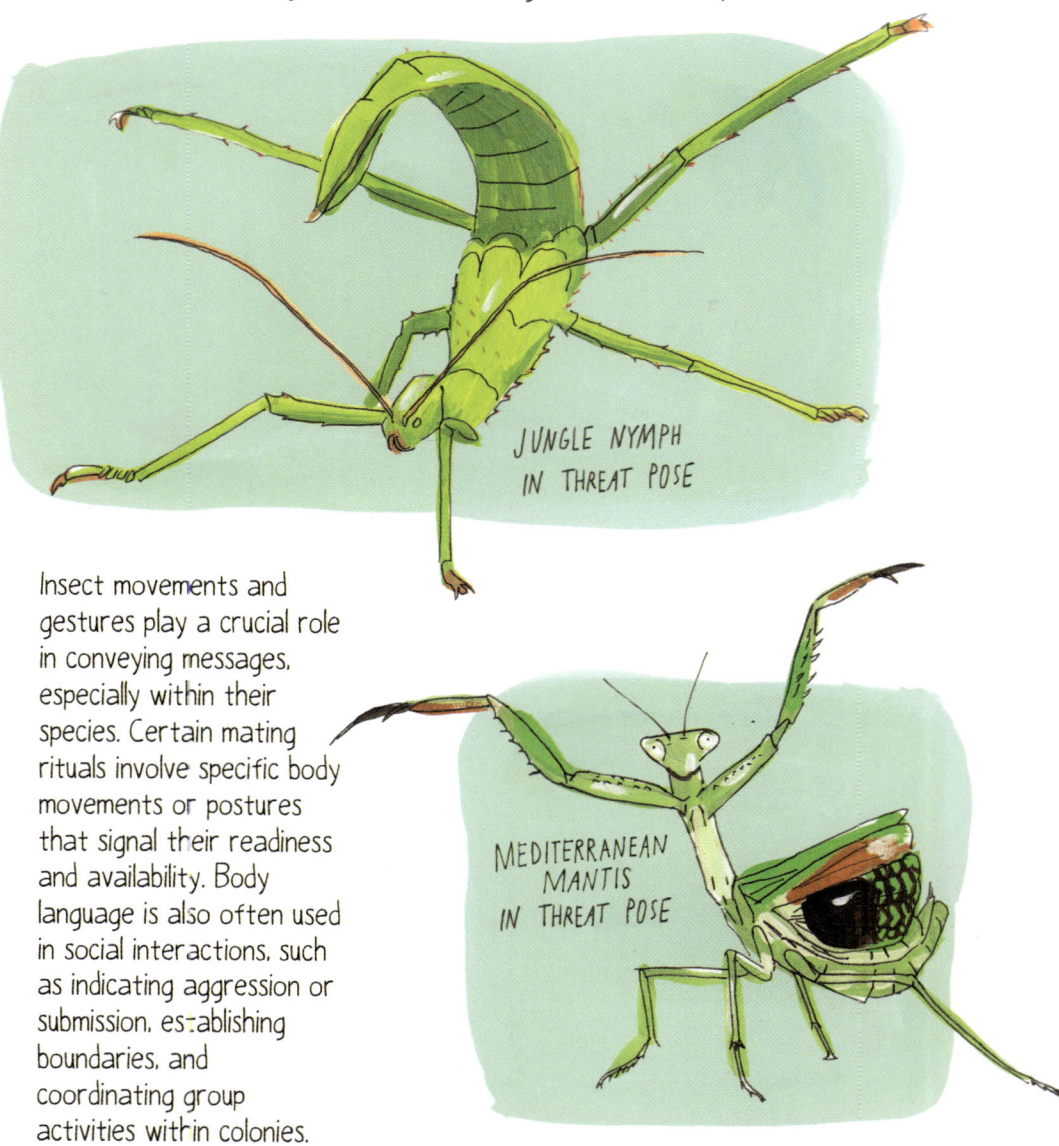

JUNGLE NYMPH
IN THREAT POSE

Insect movements and gestures play a crucial role in conveying messages, especially within their species. Certain mating rituals involve specific body movements or postures that signal their readiness and availability. Body language is also often used in social interactions, such as indicating aggression or submission, establishing boundaries, and coordinating group activities within colonies.

MEDITERRANEAN
MANTIS
IN THREAT POSE

ANATOMY OF A MANTIS

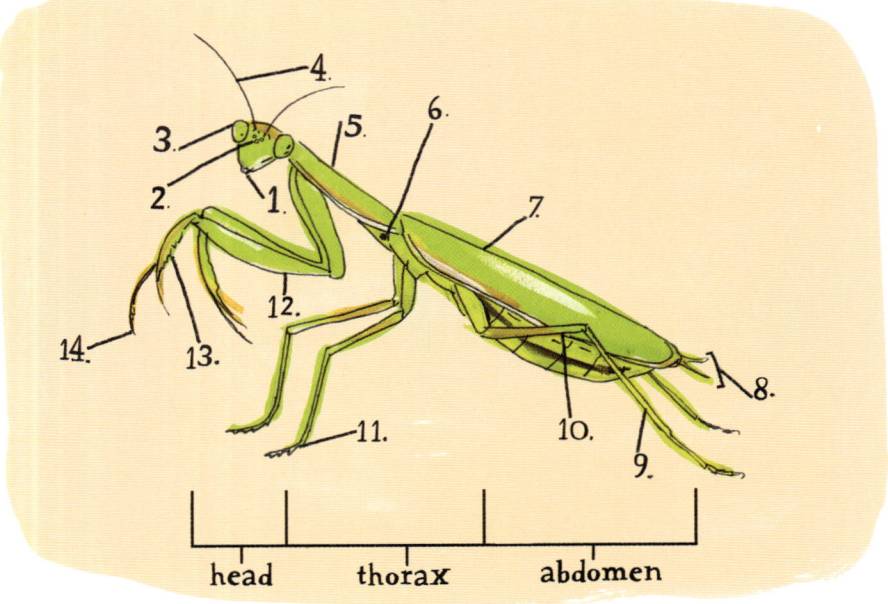

head   thorax   abdomen

1. mandible	8. cerci
2. simple eye	9. tibia
3. compound eye	10. hindleg femur
4. antenna	11. tarsus
5. pronotum	12. foreleg femur
6. ear	13. tibial spines
7. wings	14. tarsus

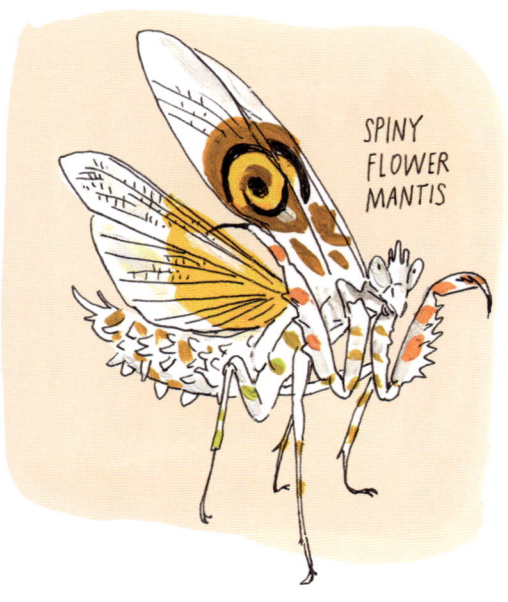

SPINY FLOWER MANTIS

Praying mantises hear through special sensors that pick up vibrations from the air. Their sensitivity to sound enhances their ability to find both mates and prey.

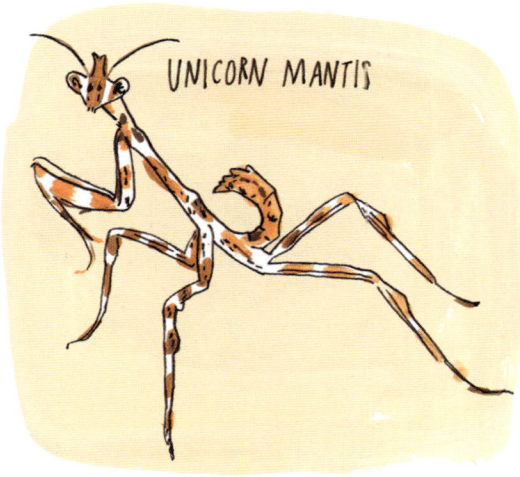

UNICORN MANTIS

Praying mantises have five eyes. Two large compound eyes detect movement and provide depth perception and a broad field of vision. Three simple eyes, called ocelli, sit in a triangle on the head. They are sensitive to changes in light.

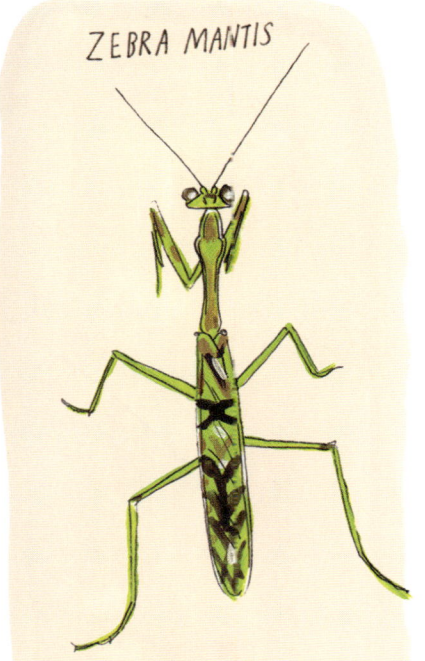

ZEBRA MANTIS

With the ability to turn their heads 180 degrees, praying mantises can quickly spot and stalk prey while staying alert to predators. Praying mantises aren't picky eaters. In addition to insects, they'll eat spiders, frogs, lizards, and even small birds.

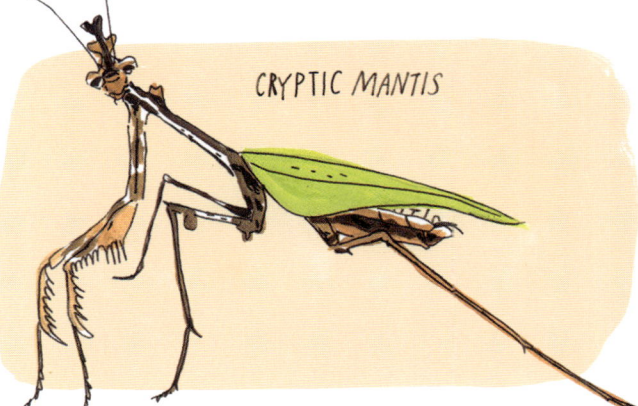

CRYPTIC MANTIS

HEAD-SHOTS

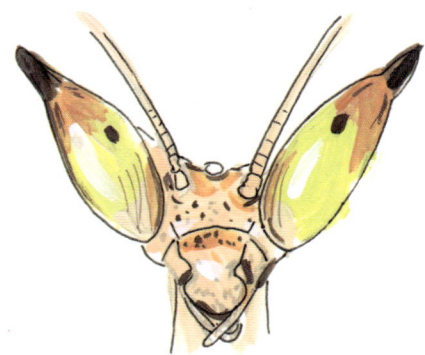

Chinese

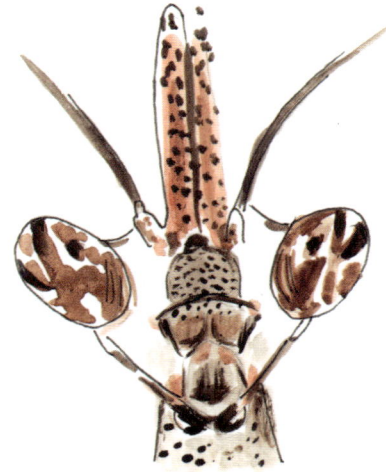

Arizona Unicorn

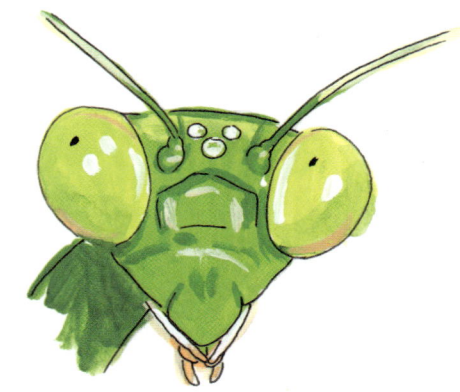

Cat

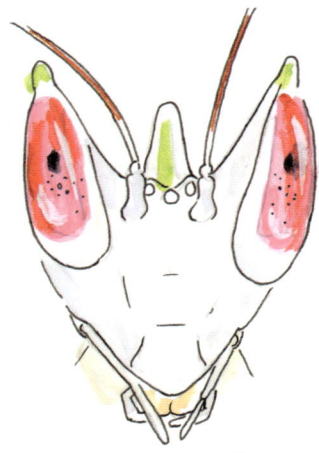

Orchid

European

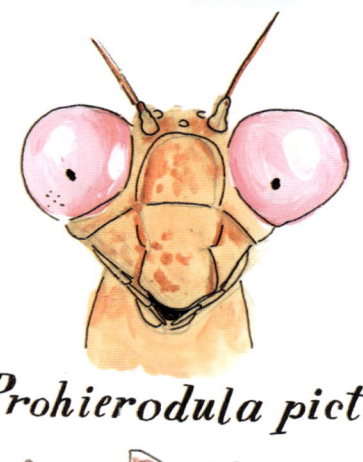

Prohierodula picta

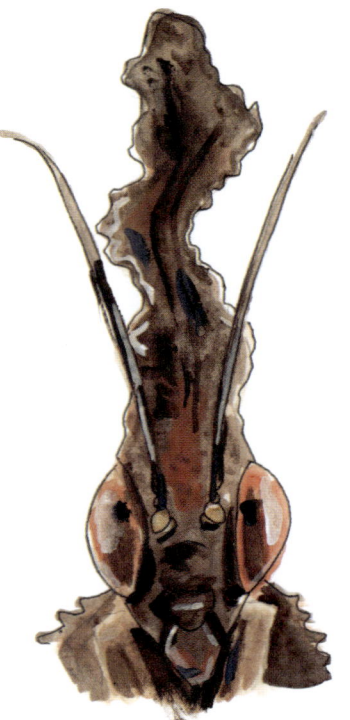

Ghost

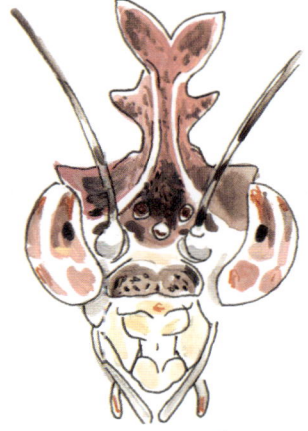

Cryptic

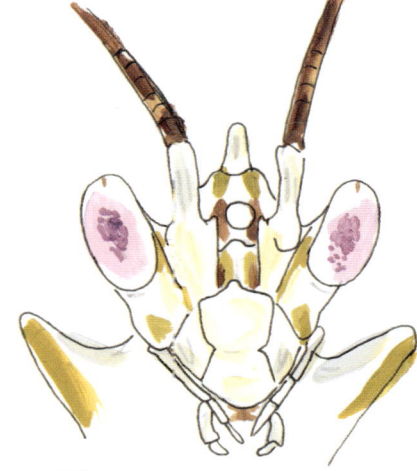

Spiny Flower

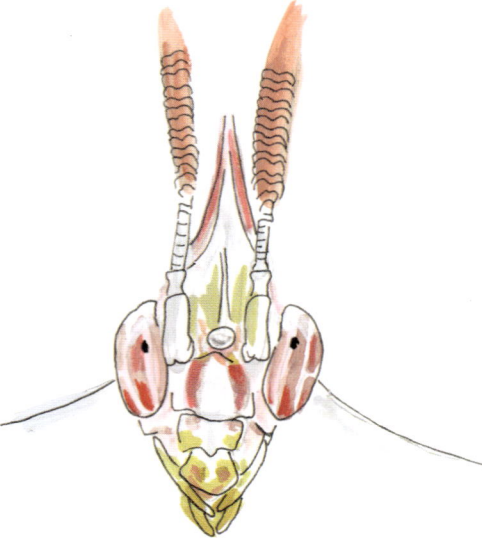

Devil's Flower

LIGHTING THE SKY

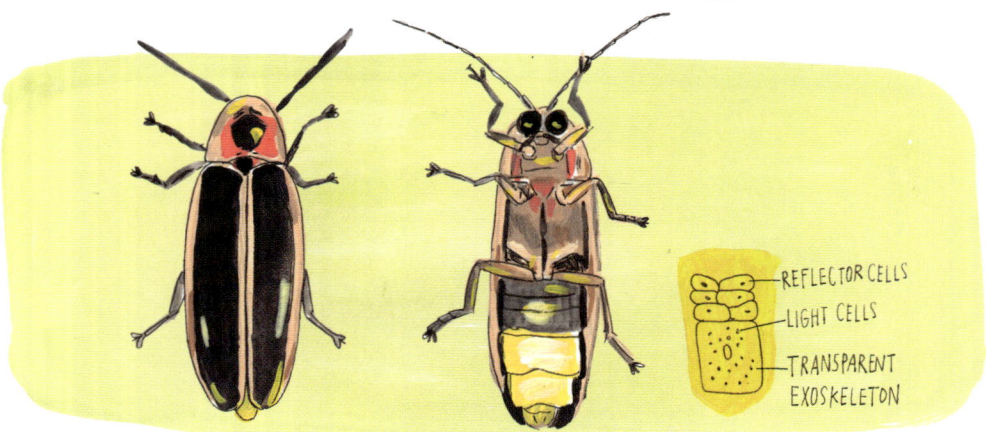

REFLECTOR CELLS
LIGHT CELLS
TRANSPARENT EXOSKELETON

Fireflies, which are soft-bodied beetles, have the remarkable ability to produce light as a form of communication. Their flashing patterns are used to attract mates. The light, called bioluminescence, is the result of a chemical reaction that takes place in a specialized abdominal organ, where they combine oxygen with a substance called luciferin.

There are around 2,000 species of firefly, each with their own unique flashing pattern.

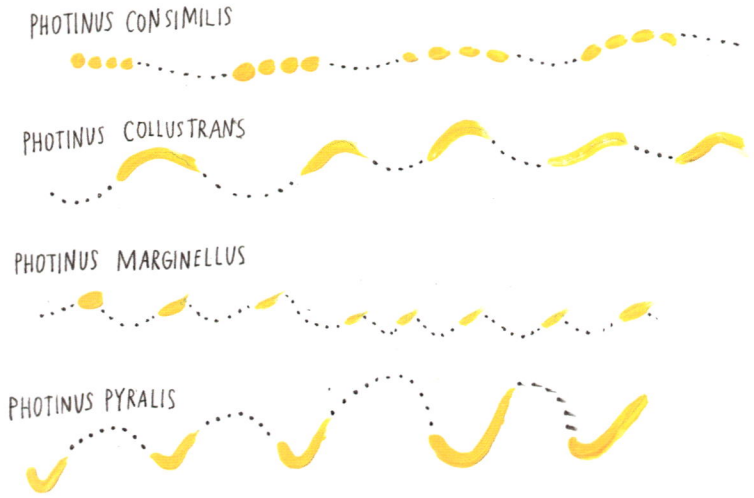

PHOTINUS CONSIMILIS

PHOTINUS COLLUSTRANS

PHOTINUS MARGINELLUS

PHOTINUS PYRALIS

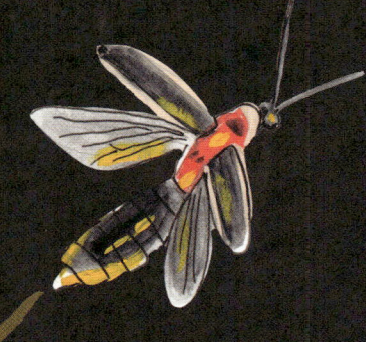

Fireflies produce "cold light," which means there's no heat generated—100 percent of the energy is emitted as light.

Fireflies have a long larval stage that can last one to two years. The adults only live a few weeks.

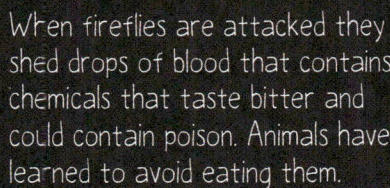

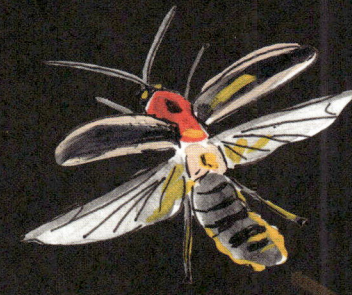

When fireflies are attacked they shed drops of blood that contains chemicals that taste bitter and could contain poison. Animals have learned to avoid eating them.

The Firefly Specialist Group of the International Union for Conservation of Nature has found at least 18 species of fireflies are at risk for extinction because of habitat destruction.

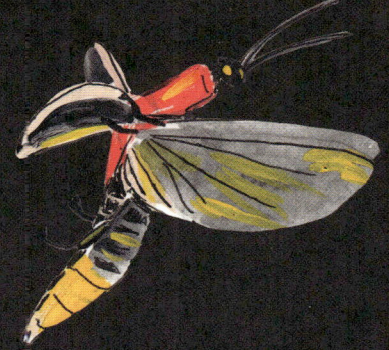

PHEROMONES

Many insects communicate using pheromones, special chemicals that create odors. The pheromones are emitted through various glands located in different parts of their bodies.

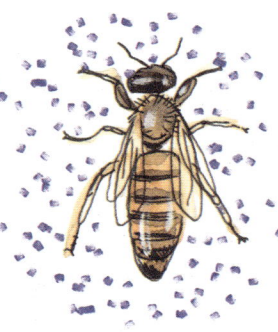

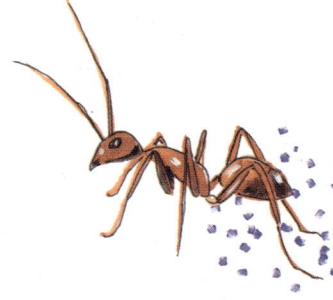

Queen honey bees produce pheromones to communicate with the hive, helping to maintain the hive's social structure.

Ants use distinct odors for trail marking and alarm signals.

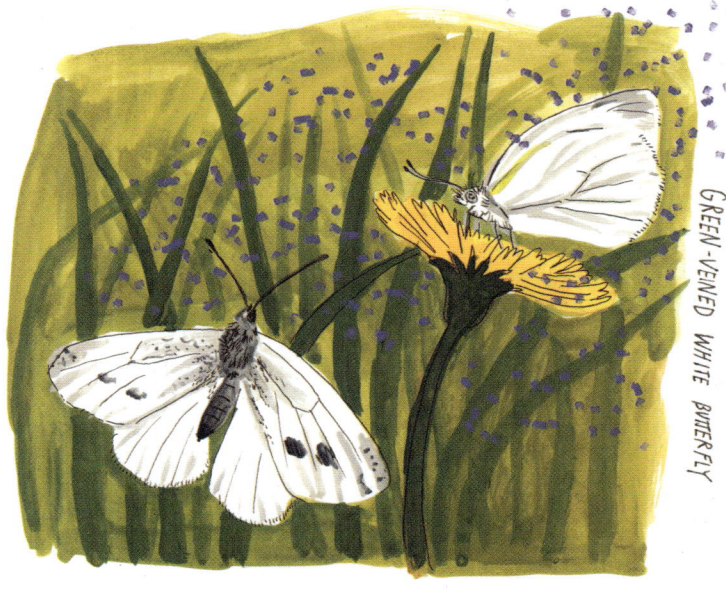

GREEN-VEINED WHITE BUTTERFLY

Some species of butterflies and moths emit pheromones to attract mates over long distances.

NOISY BUGS

Grasshoppers, locusts, crickets, and katydids produce sounds by rubbing parts of their bodies together in a process called stridulation.

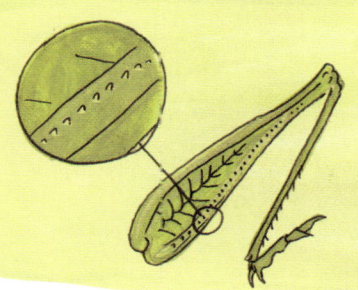

Grasshoppers have specialized pegs on the inside of their hindlegs. Their familiar buzzing noise is created when the pegs are rubbed against a ridge on one of the forewings. This sound is especially important during mating rituals or when establishing territory.

Many species of beetle, including the longhorn, stridulate by rubbing the underside of their wings against specialized ridges on their abdomen.

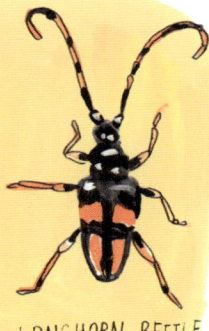

LONGHORN BEETLE

Instead of rubbing body parts together like most stridulating insects, plant bugs rapidly vibrate or tap their mouthparts against plant stems or leaves.

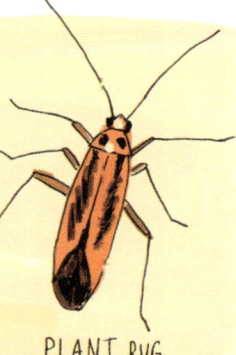

PLANT BUG

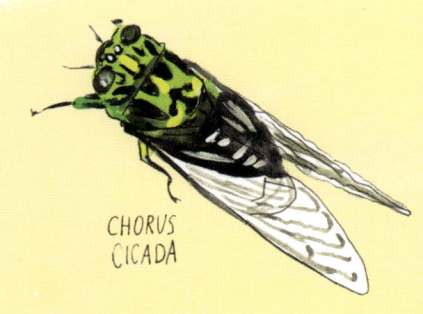

CHORUS CICADA

Cicadas are the loudest insects in the world. Some species are as loud as a chainsaw. The males make the high-pitched buzzing to attract females. There are over 3,000 different species of cicada that are either annual, appearing every year, or periodical, which emerge only once every decade or two.

COURTING AND MATING

Beyond bioluminescence, sound, and pheromones, some insects exhibit behaviors that resemble dancing, particularly during courtship rituals. These movements serve as a form of communication to attract mates or establish mating compatibility.

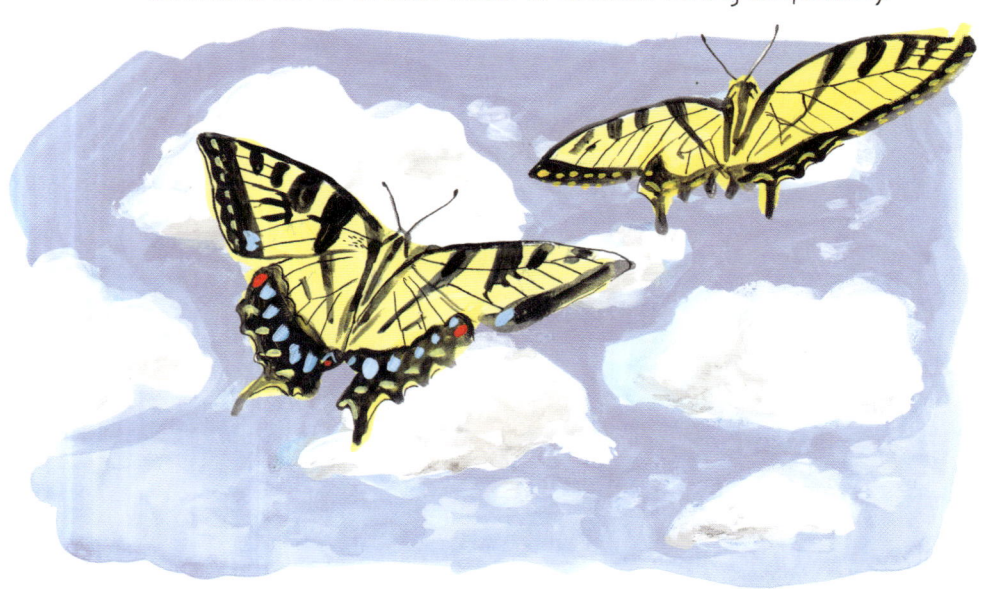

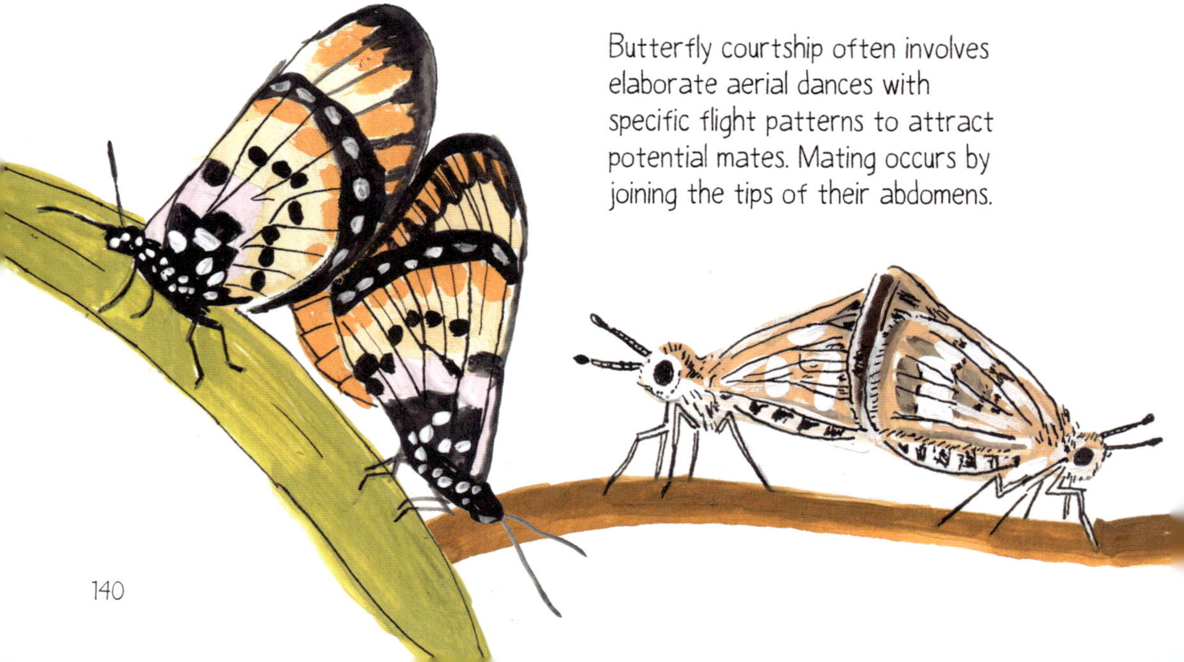

Butterfly courtship often involves elaborate aerial dances with specific flight patterns to attract potential mates. Mating occurs by joining the tips of their abdomens.

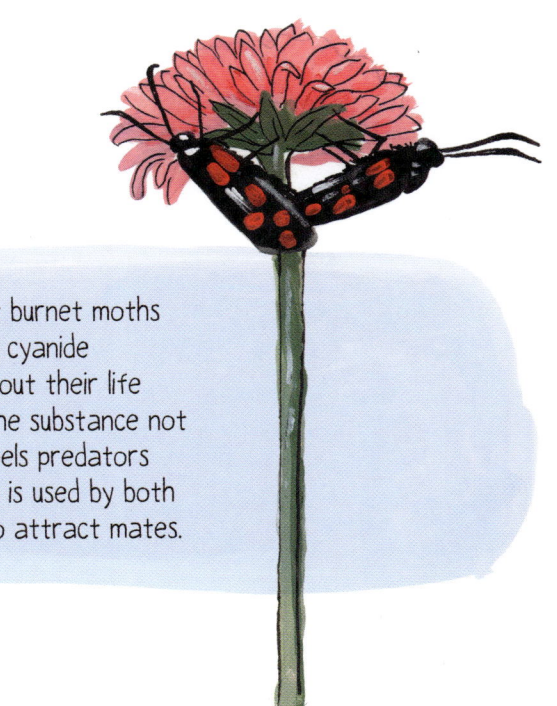

Six-spot burnet moths produce cyanide throughout their life cycle. The substance not only repels predators but also is used by both sexes to attract mates.

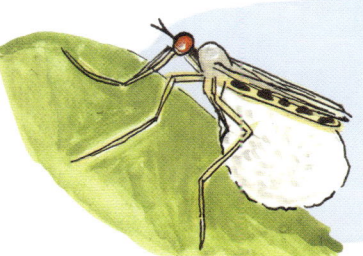

Male dance flies perform complex aerial dances to attract females, showcasing intricate flight patterns. Moreover, the male may offer a nuptial gift like a silk balloon. The balloon provides a rich protein snack for the female to eat while mating.

In the ultimate form of sacrifice, the male praying mantis is often decapitated and eaten by the female during or after mating, providing nutrients for the production of eggs.

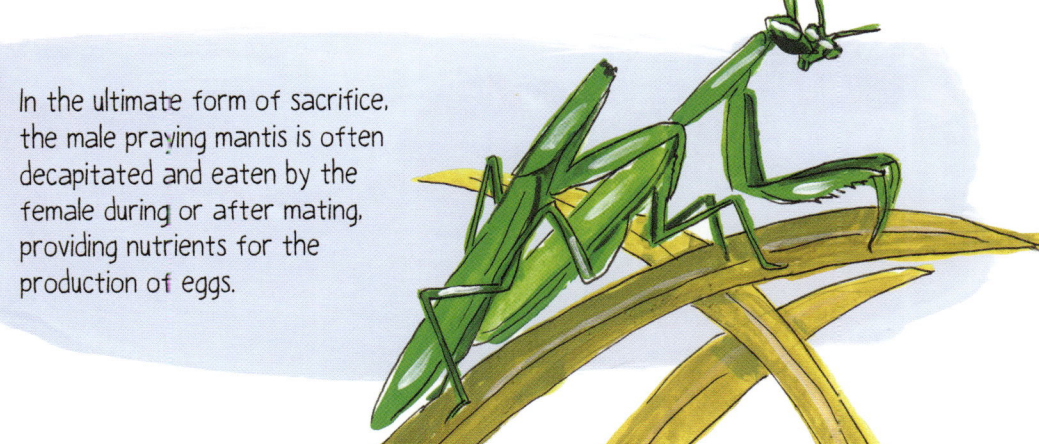

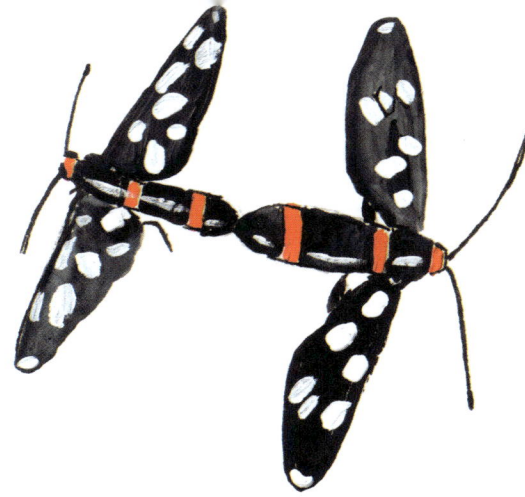

As for most creatures on the planet, mating is a crucial part of the life cycle of insects, allowing them to reproduce and ensure the survival of their species. Mating behaviors vary significantly among different insect species.

Honey bee mating typically takes place in the air away from the hive. One of several male drone bees pursuing the queen will insert his reproductive organ (endophallus) into the queen's sting chamber and ejaculate. This process is fatal for the drone, as his endophallus is torn from his body during ejaculation, causing him to die soon thereafter.

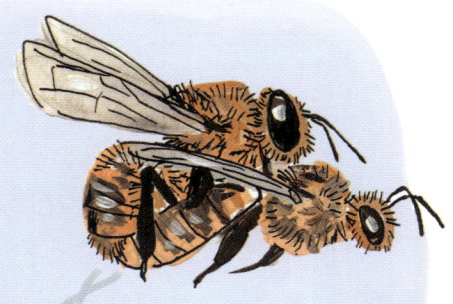

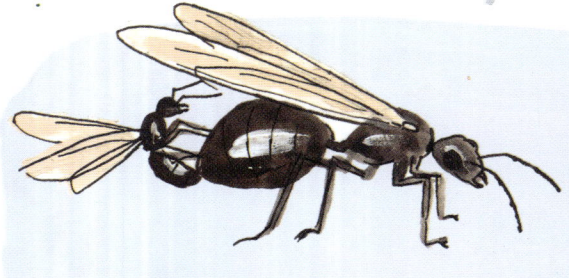

Ants have a similar mating process to honey bees, where male and female winged ants (alates) take flight in pursuit of alates from other colonies. After mating, the female establishes a new colony, where she sheds her wings and becomes queen.

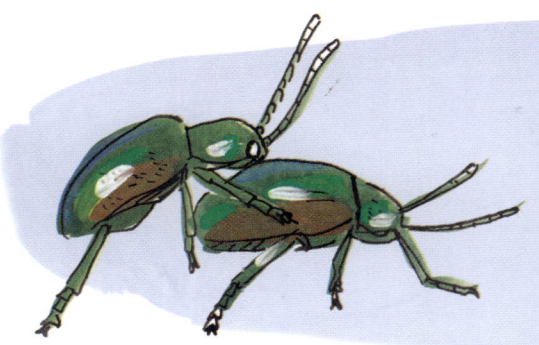

Dogbane leaf beetles mate about once a day. After mating, the male rides on the back of the female to guard her from other courting males and make sure that she uses his sperm to fertilize her eggs.

Lovebugs

A lovebug is an actual insect, not just a cute name to call your honey. These small black flies are also called honeymoon bugs because they remain coupled during mating, staying attached for several days while flying and feeding together.

These bugs are commonly found in the southeastern United States and have two swarming periods each year when dense clouds of mating pairs move together.

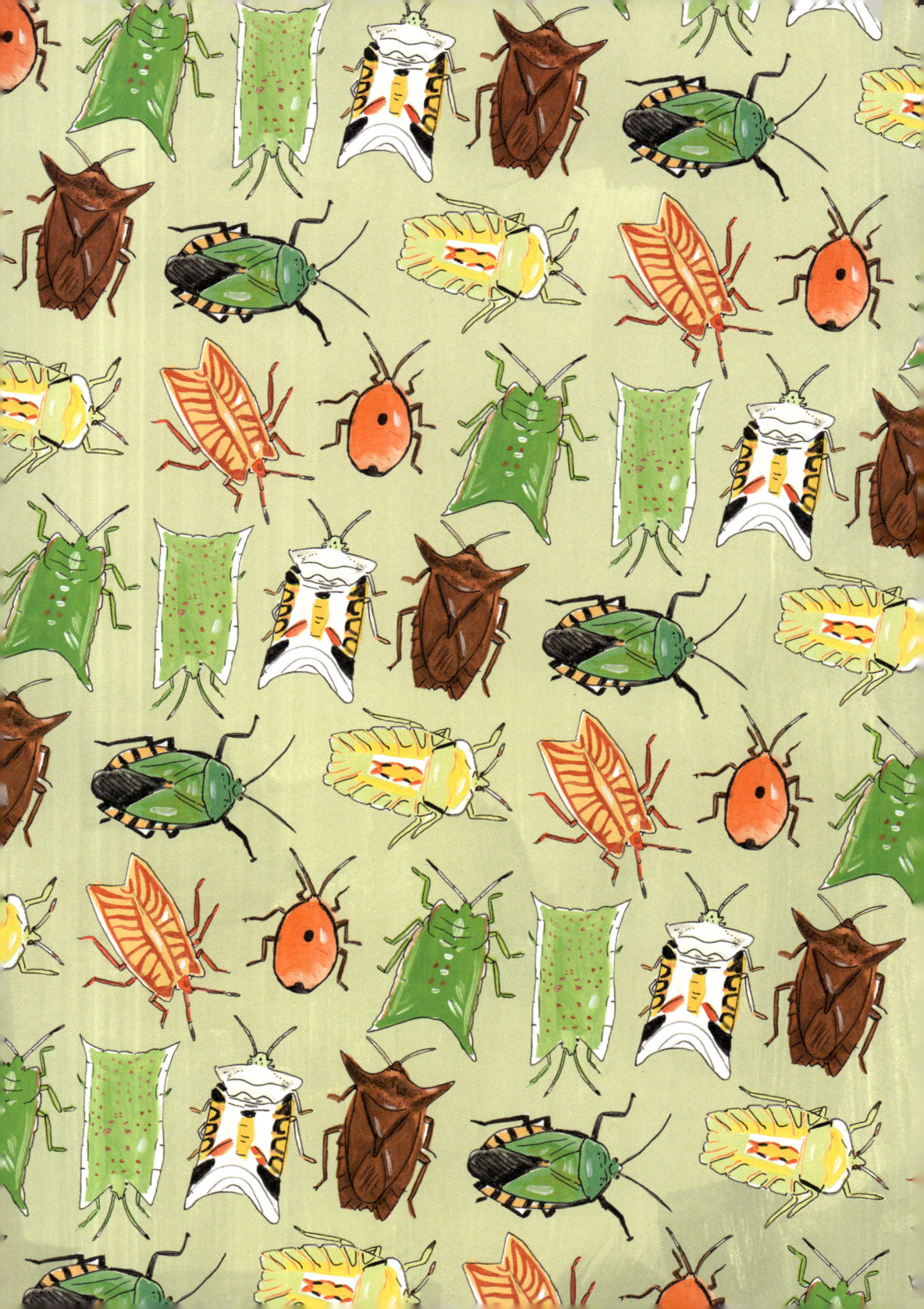

CHAPTER 6

Best in Show

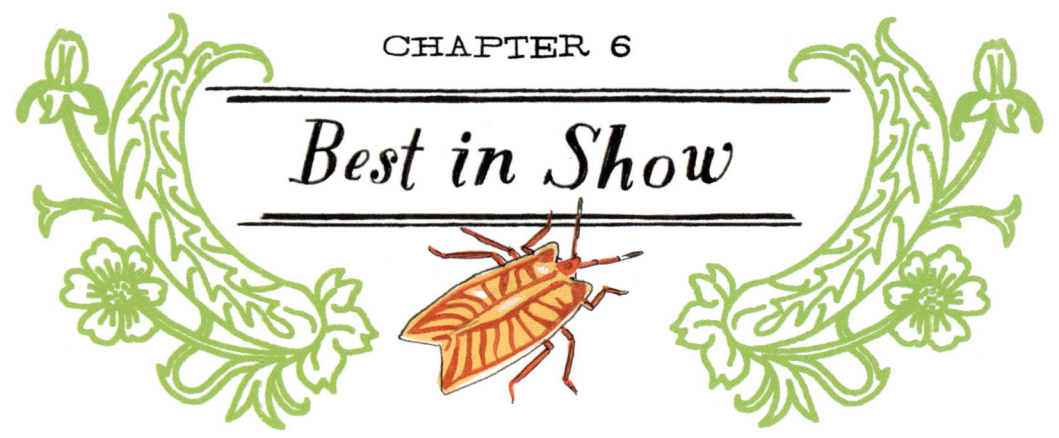

WEIRDEST

With more than 1 million known species of insects and an estimated 10 million yet to be discovered, certain species help redefine the boundaries of nature's creativity.

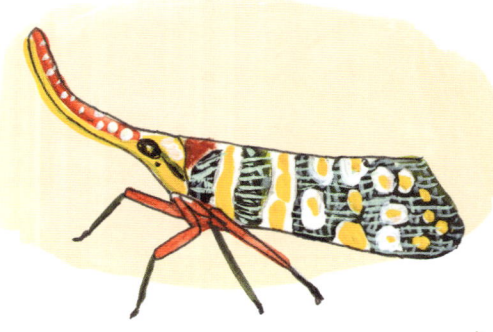

Like all planthoppers, the ASIAN LANTERNFLY uses its long proboscis to probe for plant sap.

While the male SCORPION FLY looks similar to an actual scorpion, this insect does not bite or sting humans. The curled tail is, in fact, its genitalia.

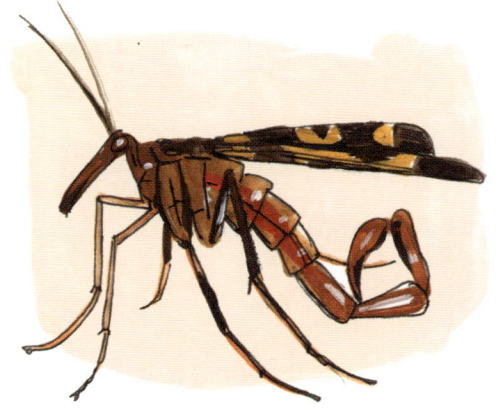

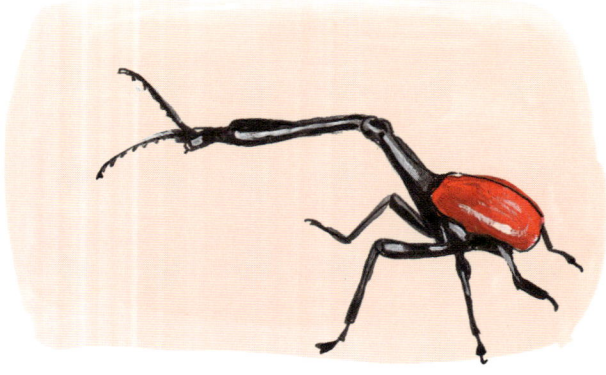

Found in Madagascar, the male GIRAFFE-NECKED WEEVIL swings its head as a weapon against other males while fighting over females.

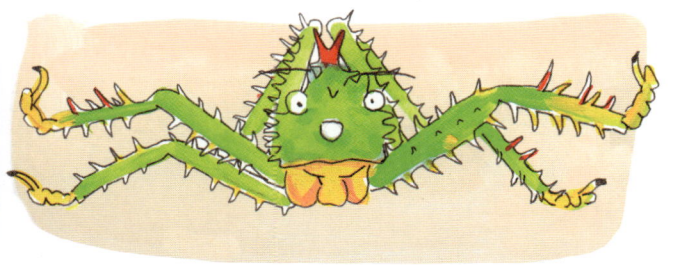

The SPINY DEVIL KATYDID uses its six prickly legs for defense. Thanks to their coloration, these nocturnal insects easily camouflage within the foliage of the tropics.

A peculiar feature of the male BAPHOMET MOTH is that it has long, inflatable tufts of hair on its abdomen. Called coremata, these tufts can be extended or retracted, and can also release pheromones to attract females.

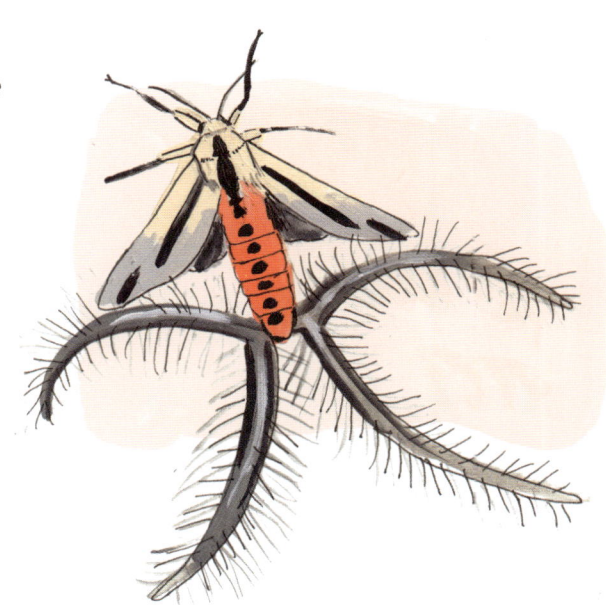

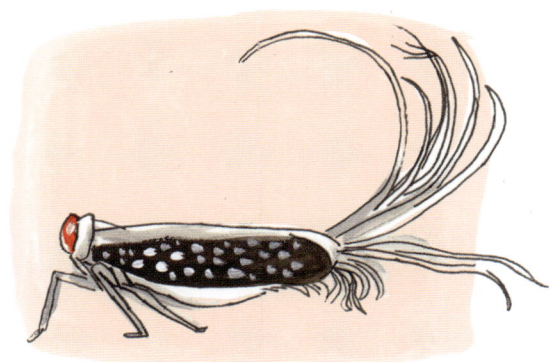

The WAX-TAIL HOPPER is native to Central and South America. It feeds on plant juices, which are converted into a type of wax that forms its feathery tail. The long tails may help protect this species from predators that end up with a mouthful of wax when they attack.

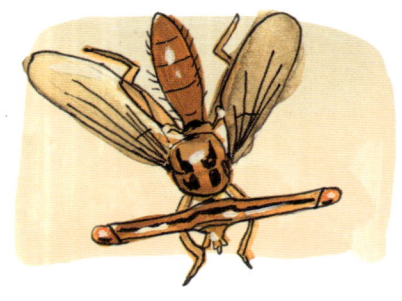

Male HAMMER-EYED FLIES have elongated eyestalks protruding from their head. Females prefer to mate with males with longer eyestalks.

Without a doubt, treehoppers and planthoppers are among the most unique-looking insects on the planet, and the BRAZILIAN TREEHOPPER is no exception. While the unusual appendages on its head are a mystery to entomologists, it is speculated that they're used to deter predators.

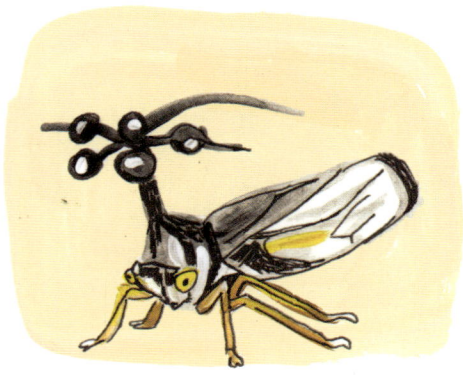

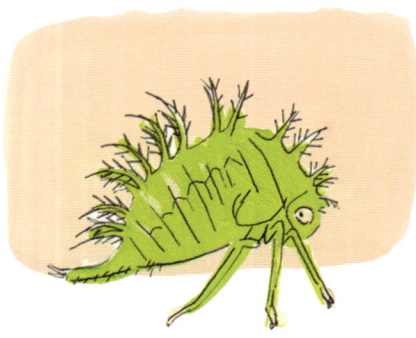

BUFFALO TREEHOPPER NYMPHS have spiny, branch-like protrusions on their bodies, helping them blend in well with surrounding plants.

The EIGHTY-EIGHT BUTTERFLY is named for its unique wing markings that look like numbers. The number 88 is considered auspicious in many cultures.

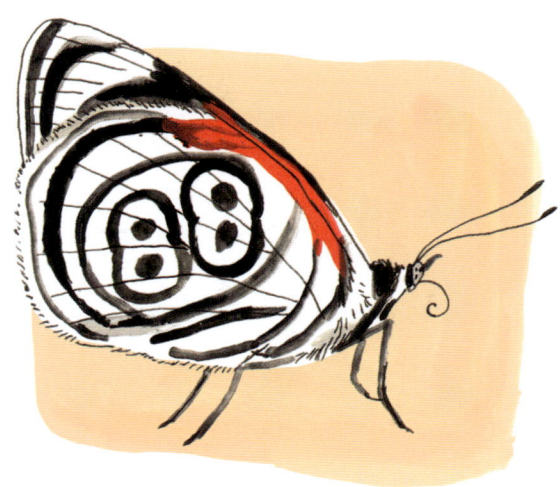

Shield Bugs

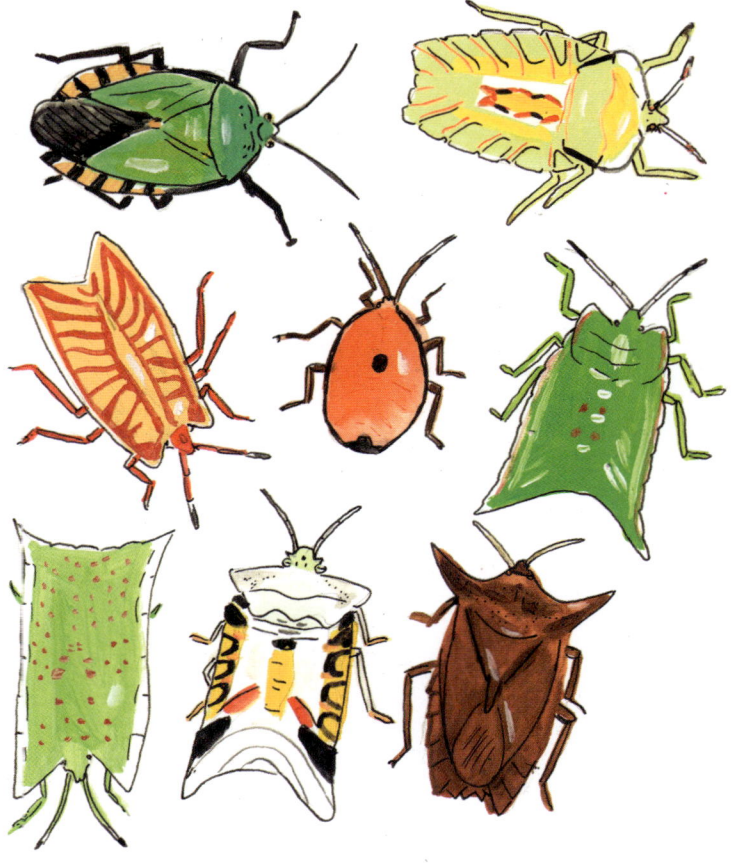

Giant shield bugs, or tessaratomids, have large, often colorful, hardened shields (known as scutella) covering their bodies, which can deter many predators. They are generally bigger than 15mm or 0.59 inches long.

Some species release a disgusting-smelling secretion from glands located on their thorax when disturbed.

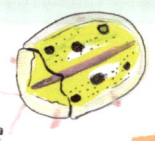

Indian green tortoise beetle

While many insects are plain black or brown, others showcase some of the most vibrant coloration and patterns among living creatures. This is particularly true for butterflies, moths, and beetles. The fancy colors on these insects are created from pigment as well as tiny and intricate ridges and valleys on the exoskeleton that absorb and reflect different wavelengths of light.

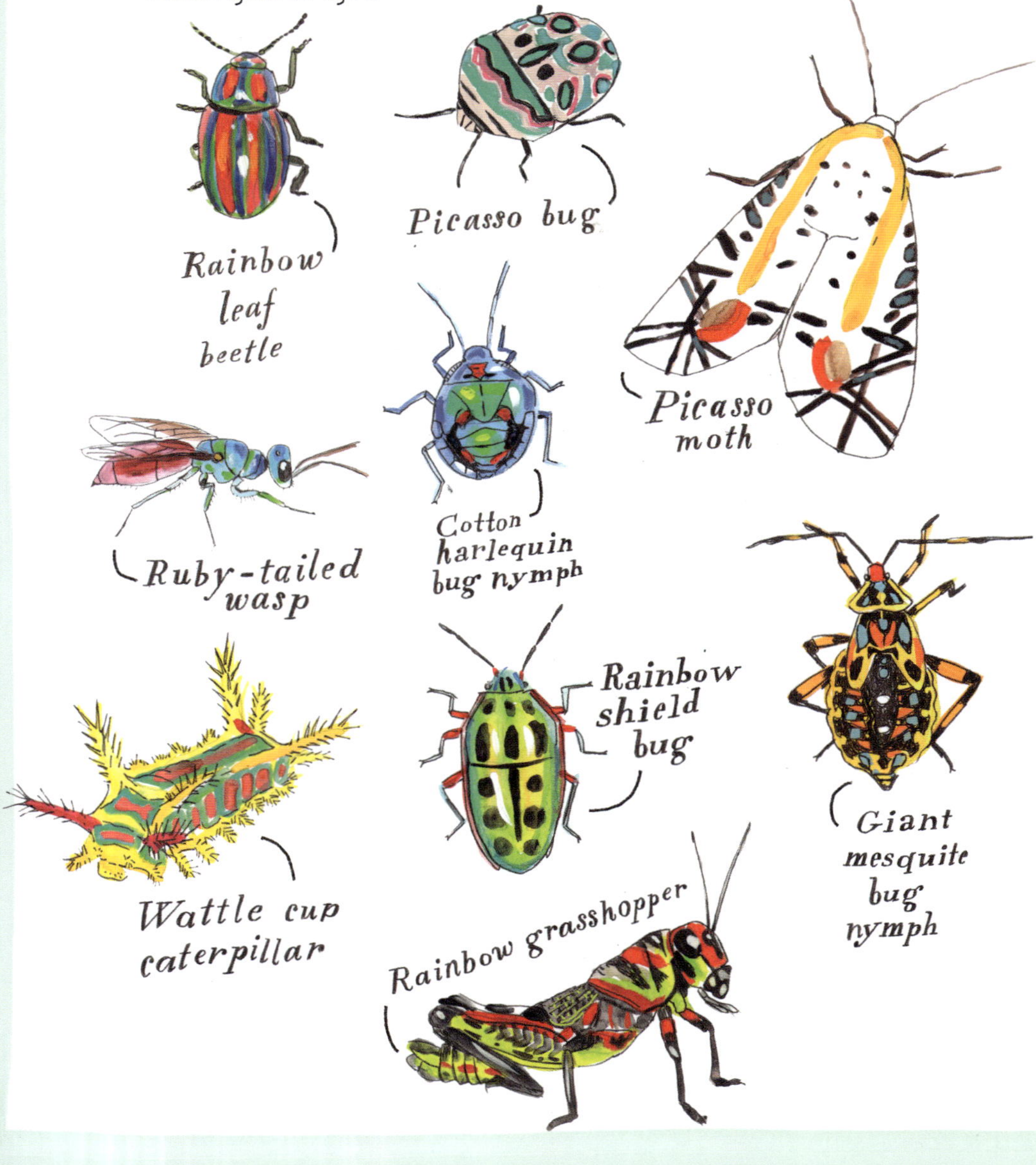

Rainbow leaf beetle

Picasso bug

Picasso moth

Ruby-tailed wasp

Cotton harlequin bug nymph

Rainbow shield bug

Giant mesquite bug nymph

Wattle cup caterpillar

Rainbow grasshopper

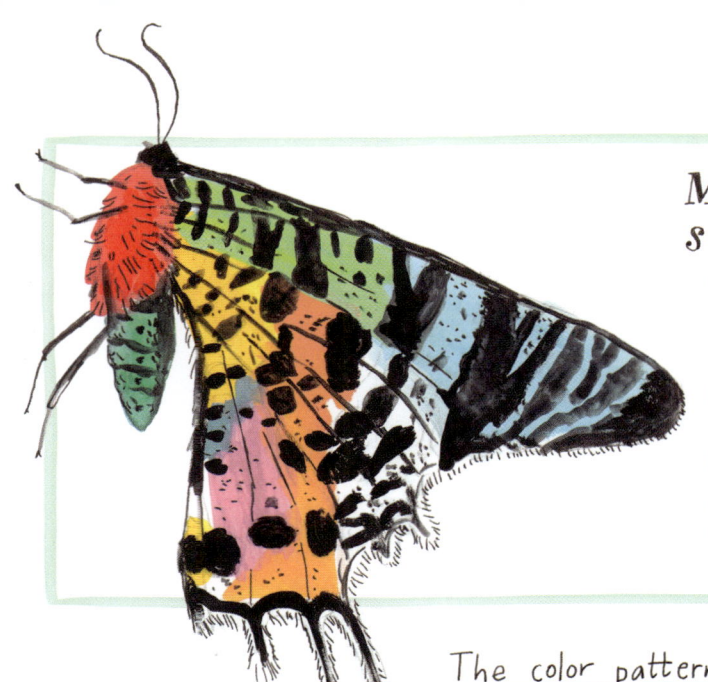

Madagascar sunset moth

Often mistaken for a butterfly, this day-flying moth is famous worldwide for its beauty, though it is only found in Madagascar.

The color patterns on insects have evolved to attract mates, hide from predators, establish territories, and even ambush prey.

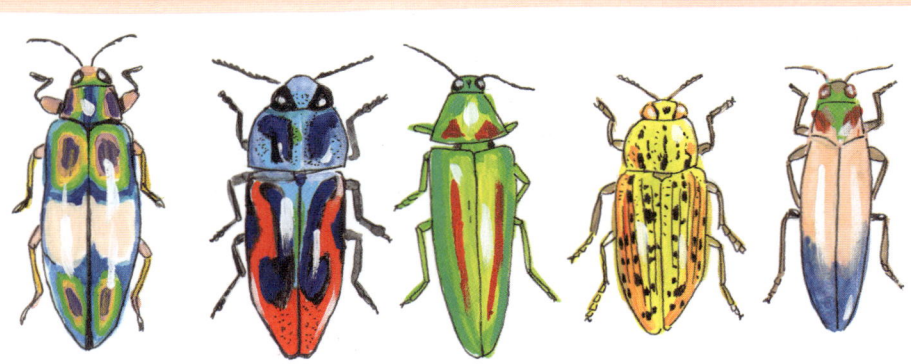

Jewel beetles

The microscopic structures on jewel beetles' exoskeltons reflect light, creating metallic, shimmering coloration. Jewel beetles are found worldwide, with 15,000 known species. They are appreciated in the art and jewelry of many cultures.

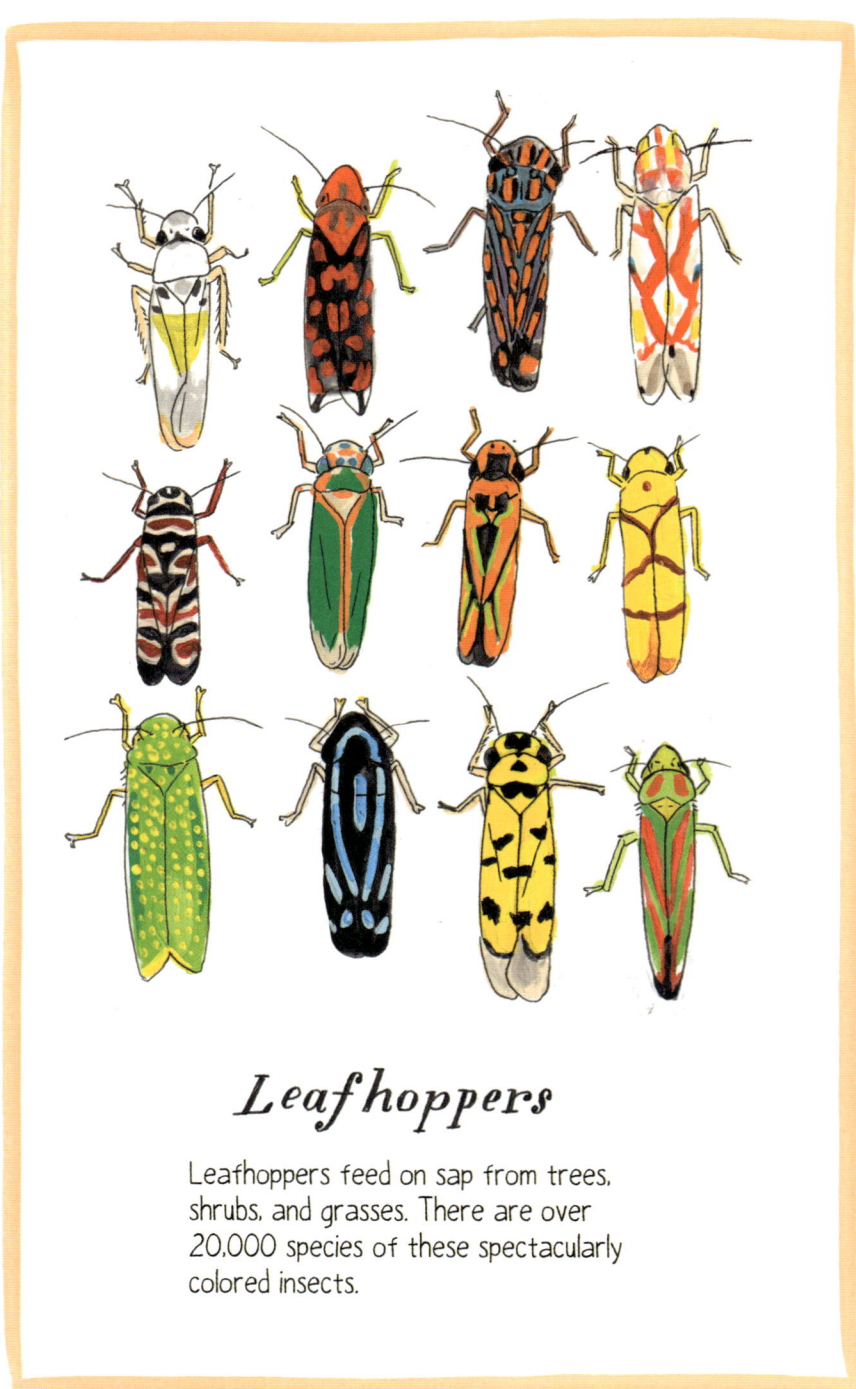

Leafhoppers

Leafhoppers feed on sap from trees, shrubs, and grasses. There are over 20,000 species of these spectacularly colored insects.

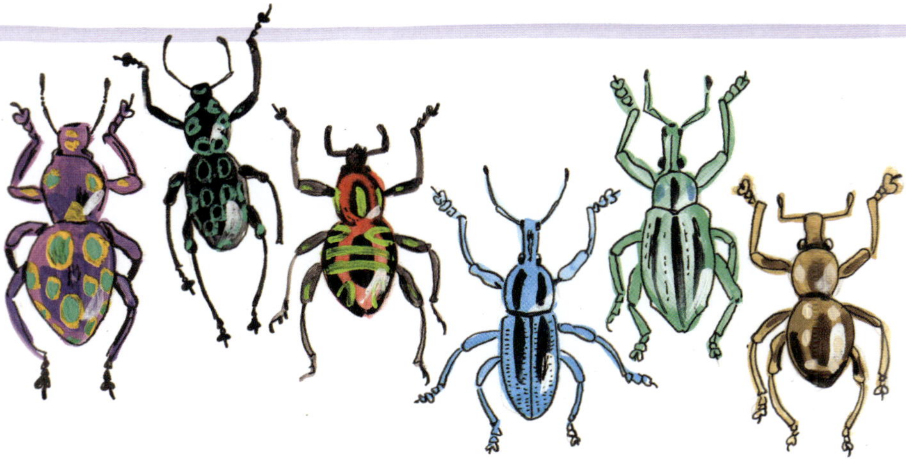

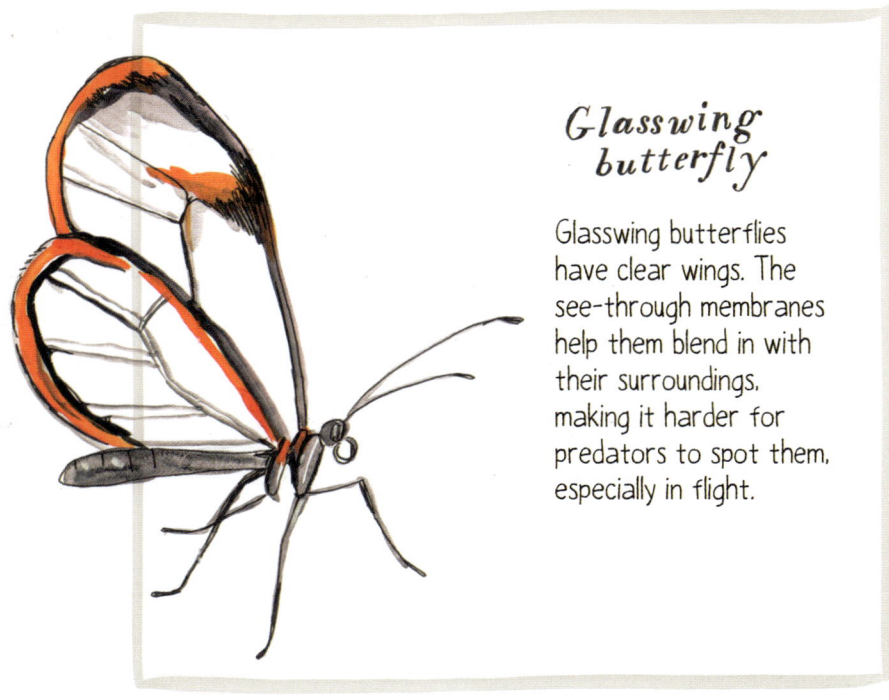

Easter egg weevils

Easter egg weevils are mostly found on Southeast Asian islands. Their colorfully patterned bodies signal to predators that they are not worth eating.

Glasswing butterfly

Glasswing butterflies have clear wings. The see-through membranes help them blend in with their surroundings, making it harder for predators to spot them, especially in flight.

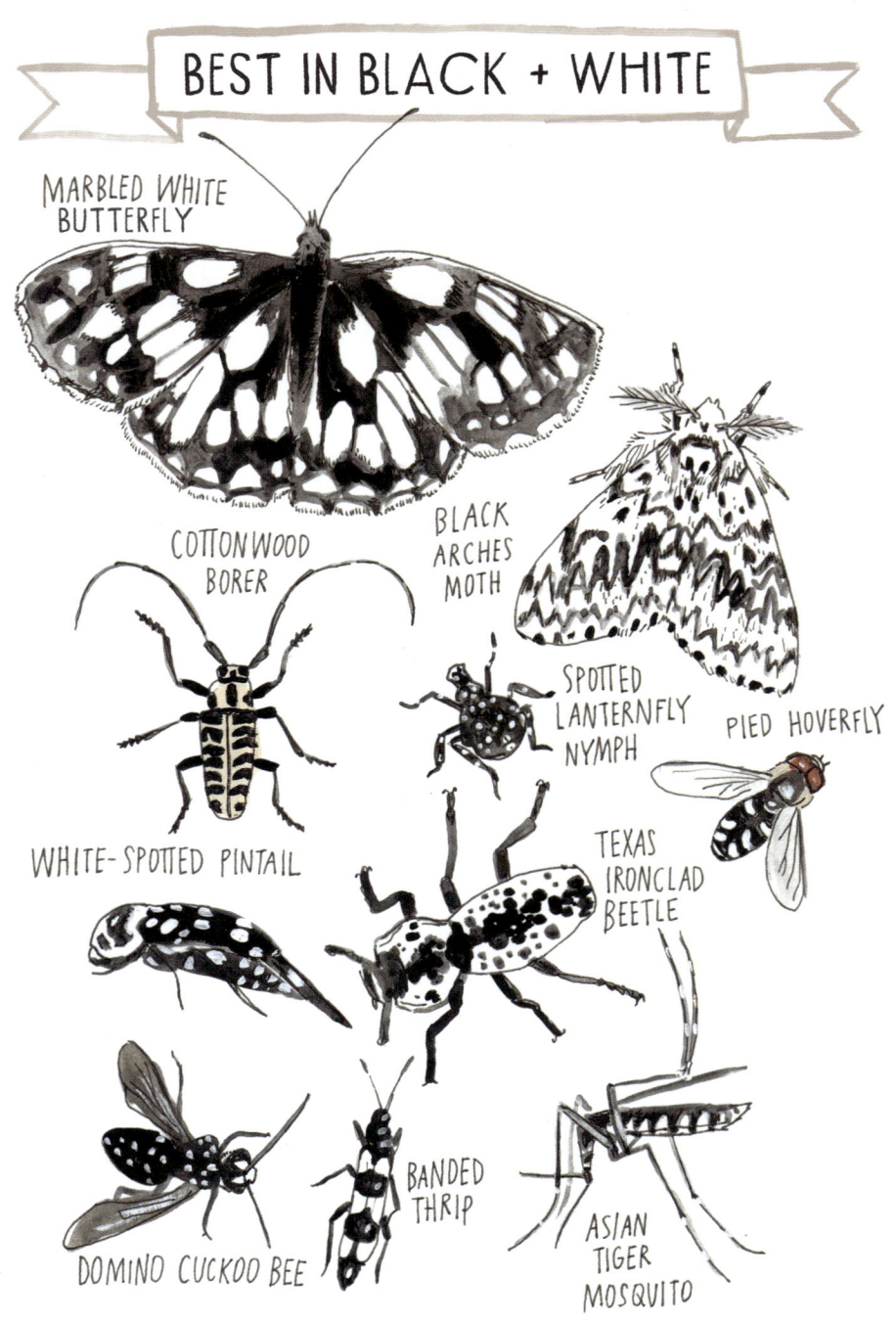

BEST IN BLACK + WHITE

MARBLED WHITE BUTTERFLY

COTTONWOOD BORER

BLACK ARCHES MOTH

SPOTTED LANTERNFLY NYMPH

PIED HOVERFLY

WHITE-SPOTTED PINTAIL

TEXAS IRONCLAD BEETLE

DOMINO CUCKOO BEE

BANDED THRIP

ASIAN TIGER MOSQUITO

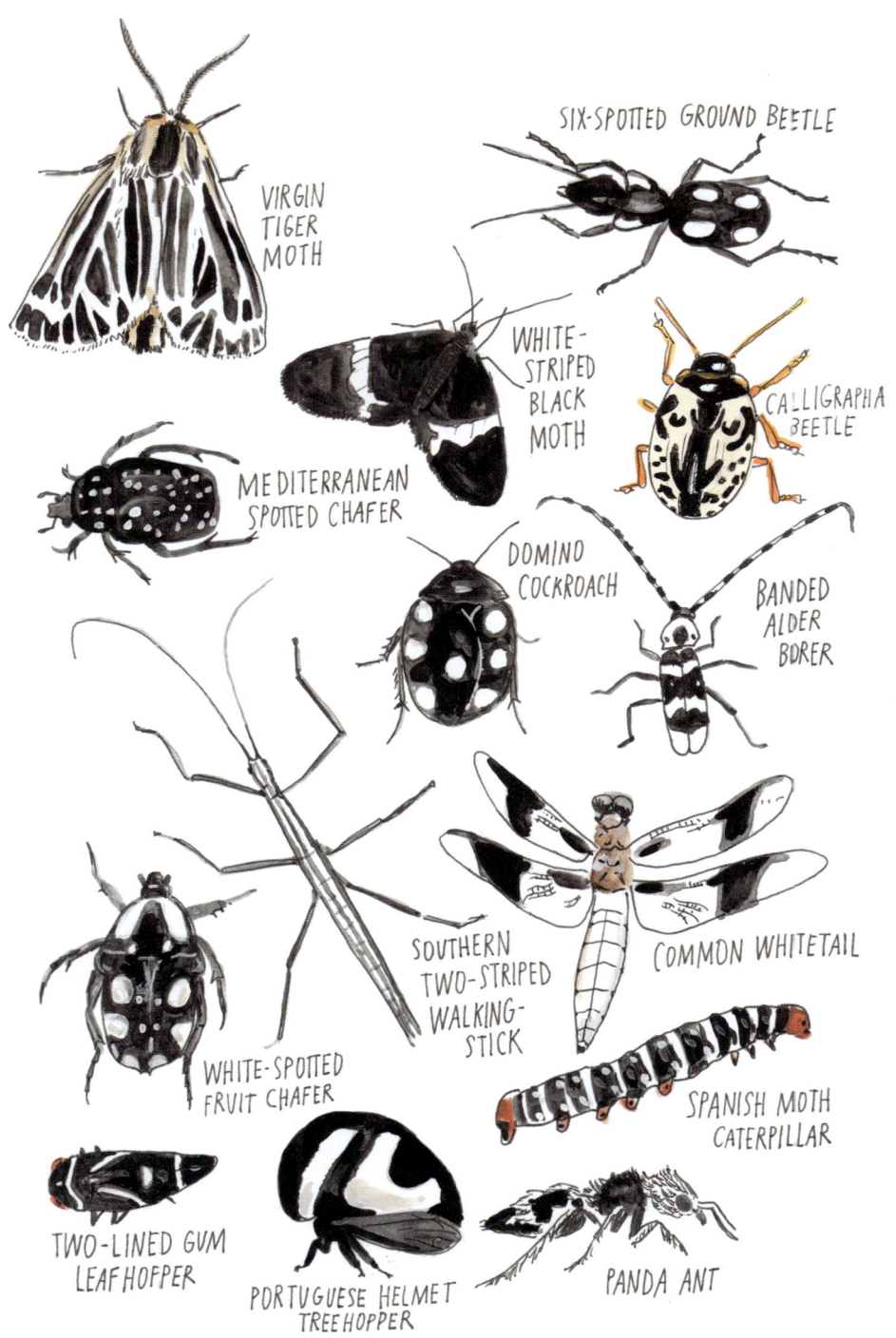

VIRGIN TIGER MOTH

SIX-SPOTTED GROUND BEETLE

WHITE-STRIPED BLACK MOTH

CALLIGRAPHA BEETLE

MEDITERRANEAN SPOTTED CHAFER

DOMINO COCKROACH

BANDED ALDER BORER

SOUTHERN TWO-STRIPED WALKING-STICK

COMMON WHITETAIL

WHITE-SPOTTED FRUIT CHAFER

SPANISH MOTH CATERPILLAR

TWO-LINED GUM LEAFHOPPER

PORTUGUESE HELMET TREEHOPPER

PANDA ANT

LONGEST

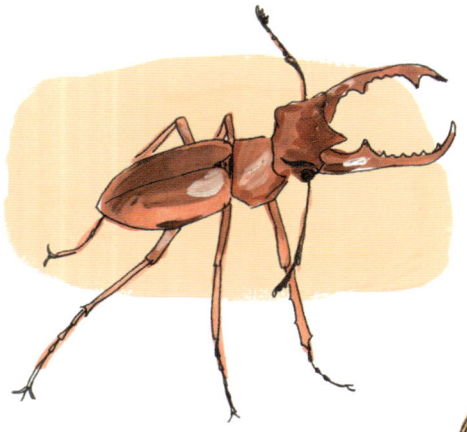

The *MANDIBLES* of the elk beetle are almost an inch long, making their total body length about 2.5 inches.

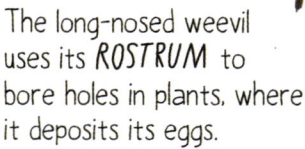

The long-nosed weevil uses its *ROSTRUM* to bore holes in plants, where it deposits its eggs.

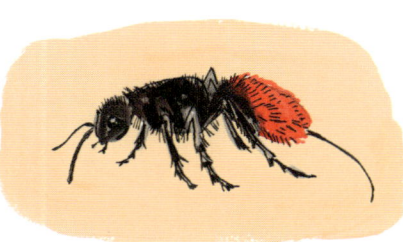

The velvet ant, sometimes called a cow killer because of the strength of its sting, is actually a wasp. The female's *STINGER* is half an inch long.

Wallace's sphinx moth has a *PROBOSCIS* that is about 8.5 inches long, an adaptation that helps it draw nectar from flowers like Darwin's orchid.

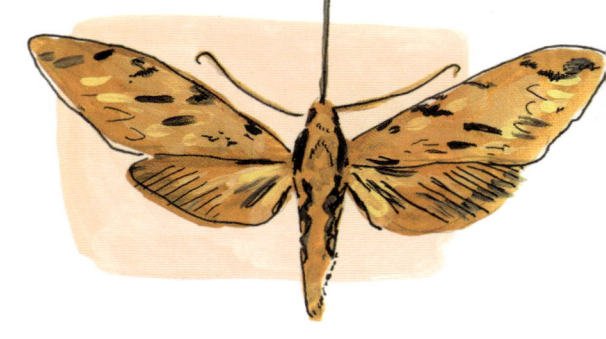

FASTEST

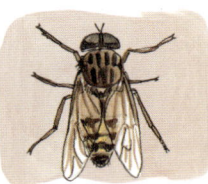

The common horsefly can reach estimated flying speeds of 90 mph in the right conditions.

LOUDEST

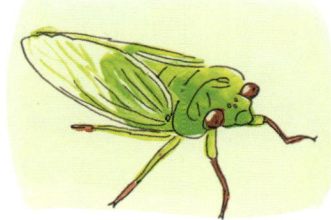

Greengrocer cicadas of Australia are believed to be one of the loudest insects on the planet. The sound they make can reach up to 120 decibels, which is about the same volume as heavy machinery or standing next to a blaring car horn.

STRONGEST

The ironclad beetle can withstand a force of nearly 40,000 times its body weight. This is equivalent to a human enduring a weight of nearly eight million pounds.

The elytra blades (forewings) on the ironclad beetle are shaped like elliptical jigsaw puzzle pieces, which are stronger than the triangular or semicircular blades found in other beetles.

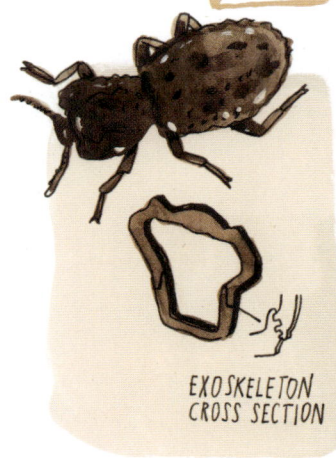

EXOSKELETON CROSS SECTION

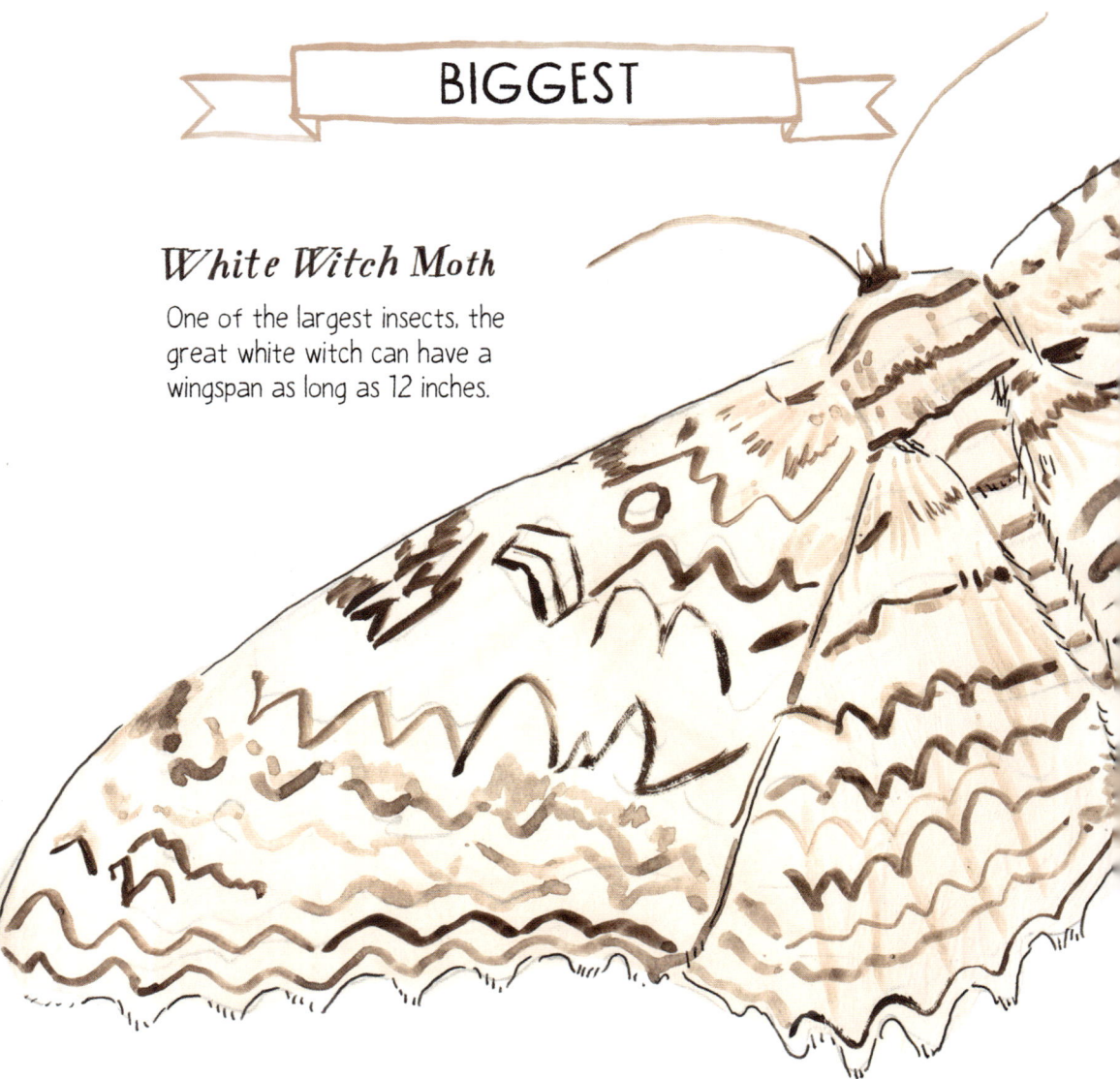

White Witch Moth

One of the largest insects, the great white witch can have a wingspan as long as 12 inches.

The insects in this section are at ACTUAL SIZE. Compare them to the size of your hand - they're really big!

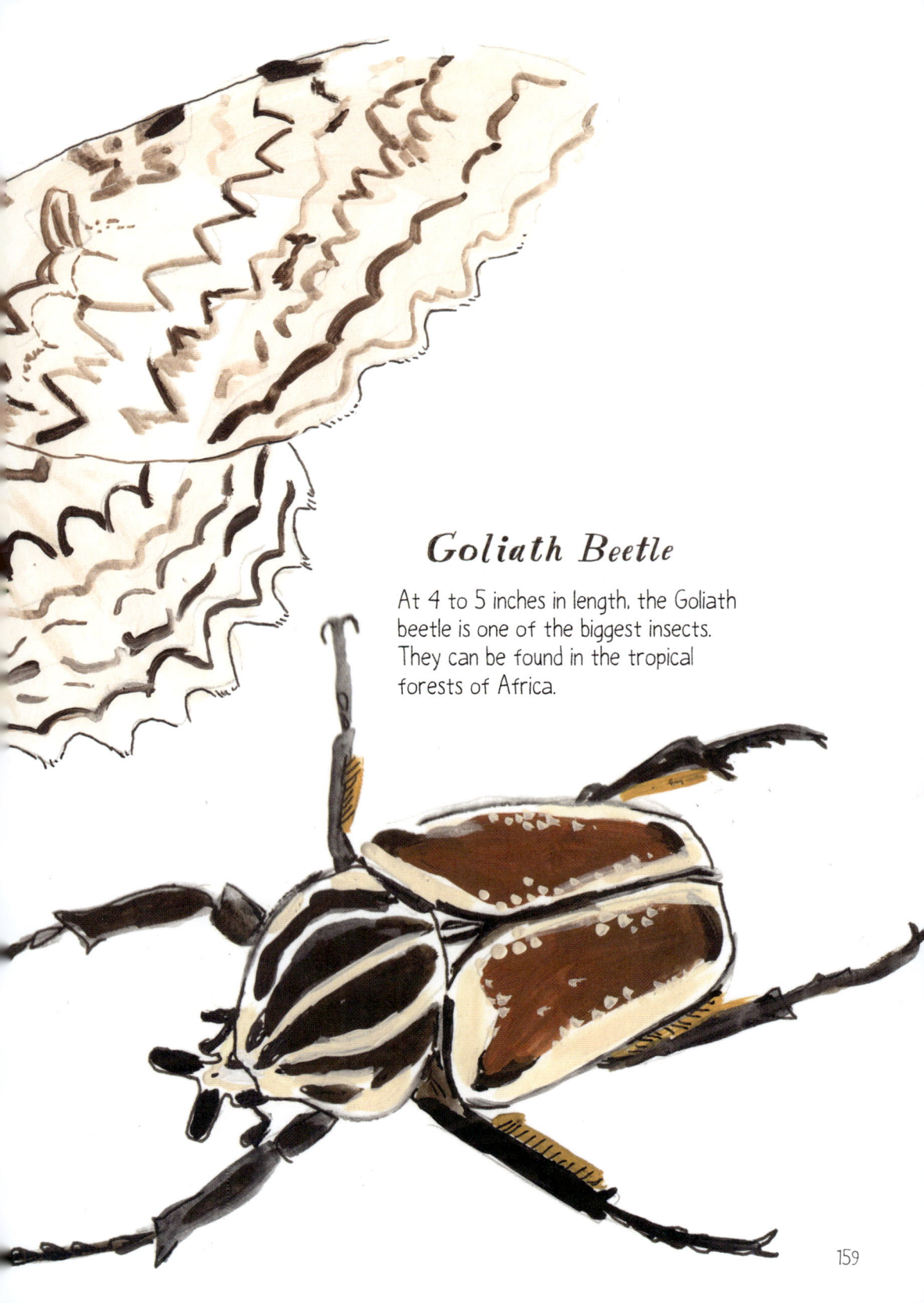

Goliath Beetle

At 4 to 5 inches in length, the Goliath beetle is one of the biggest insects. They can be found in the tropical forests of Africa.

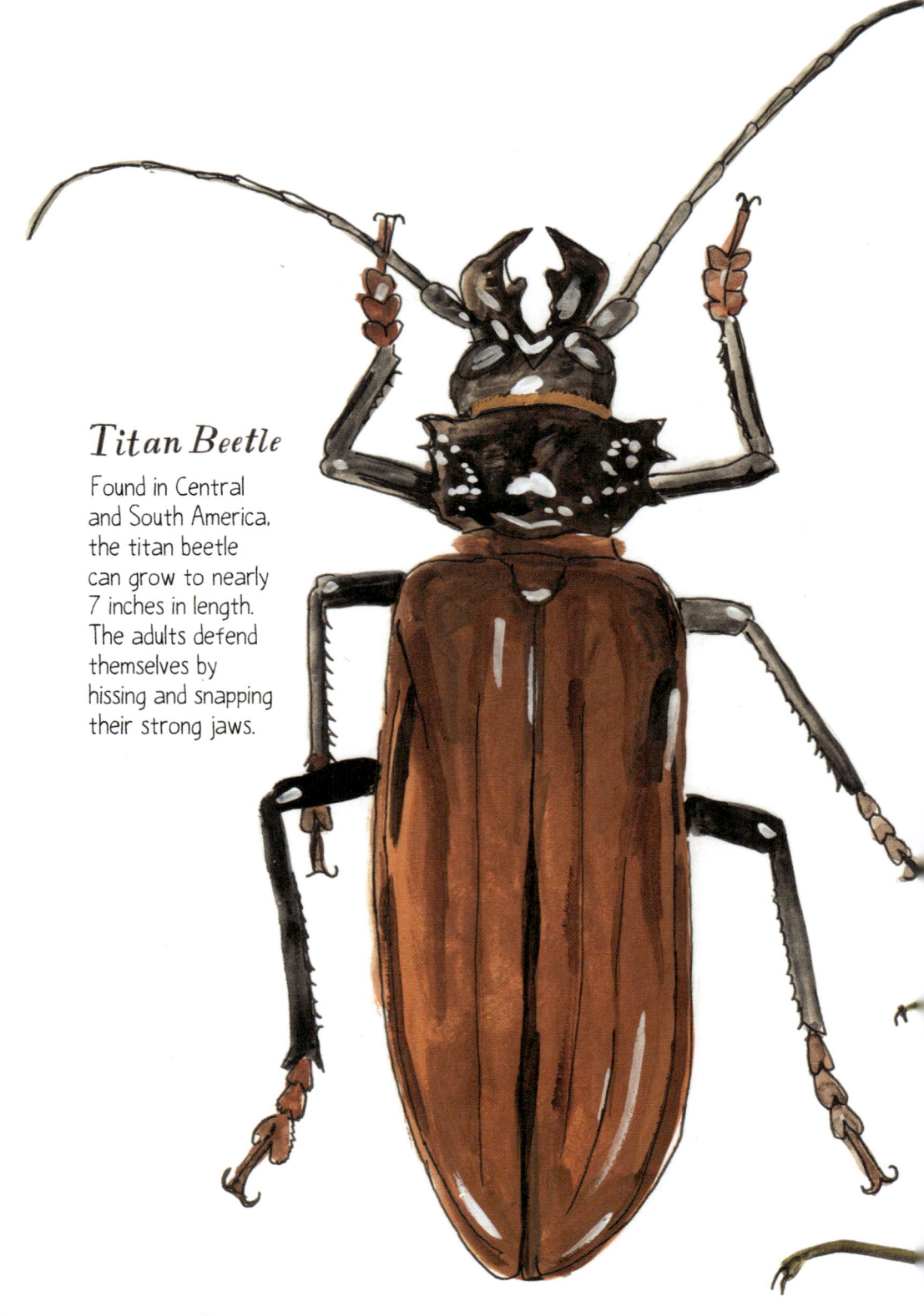

Titan Beetle

Found in Central
and South America,
the titan beetle
can grow to nearly
7 inches in length.
The adults defend
themselves by
hissing and snapping
their strong jaws.

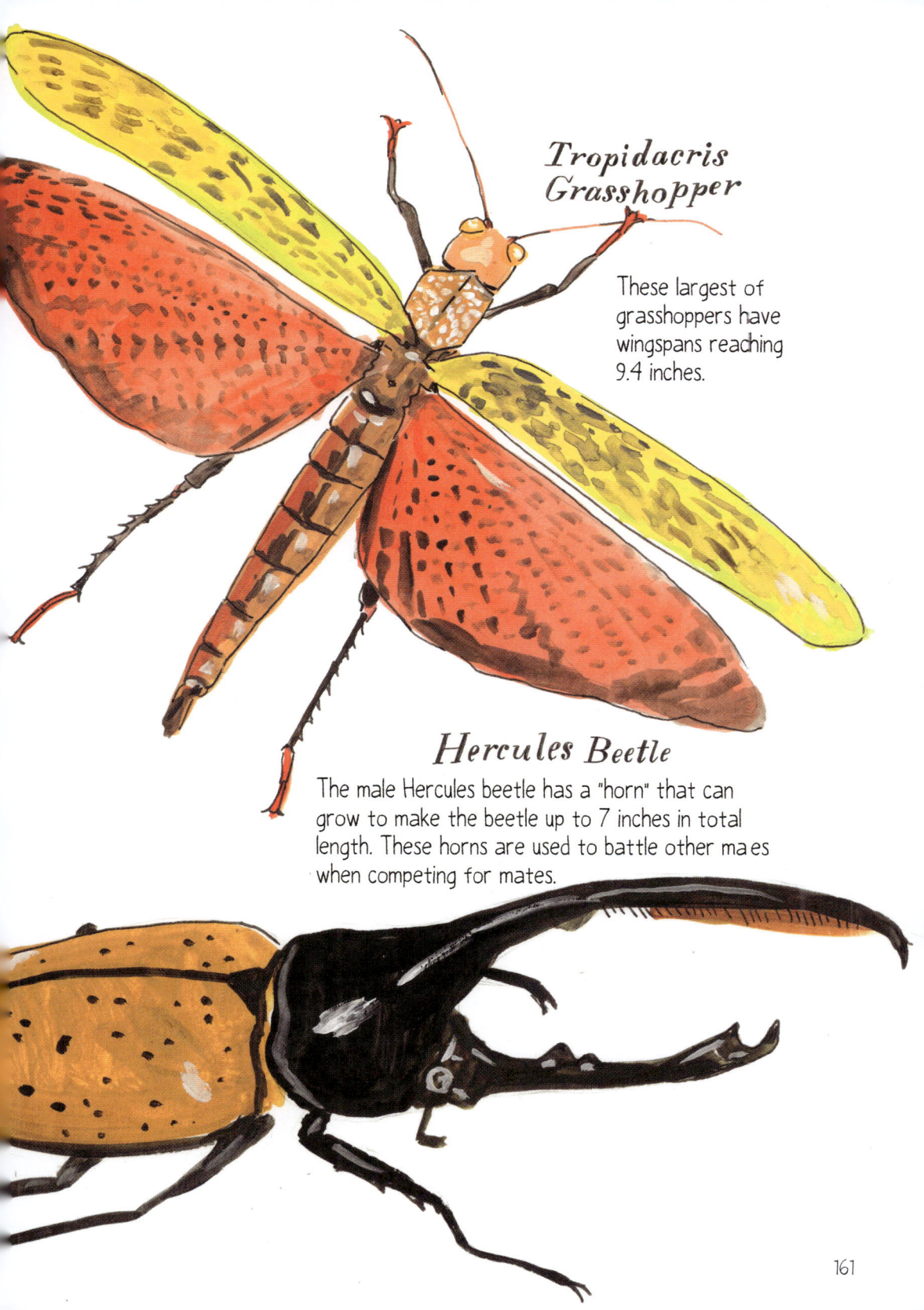

Tropidacris Grasshopper

These largest of grasshoppers have wingspans reaching 9.4 inches.

Hercules Beetle

The male Hercules beetle has a "horn" that can grow to make the beetle up to 7 inches in total length. These horns are used to battle other maes when competing for mates.

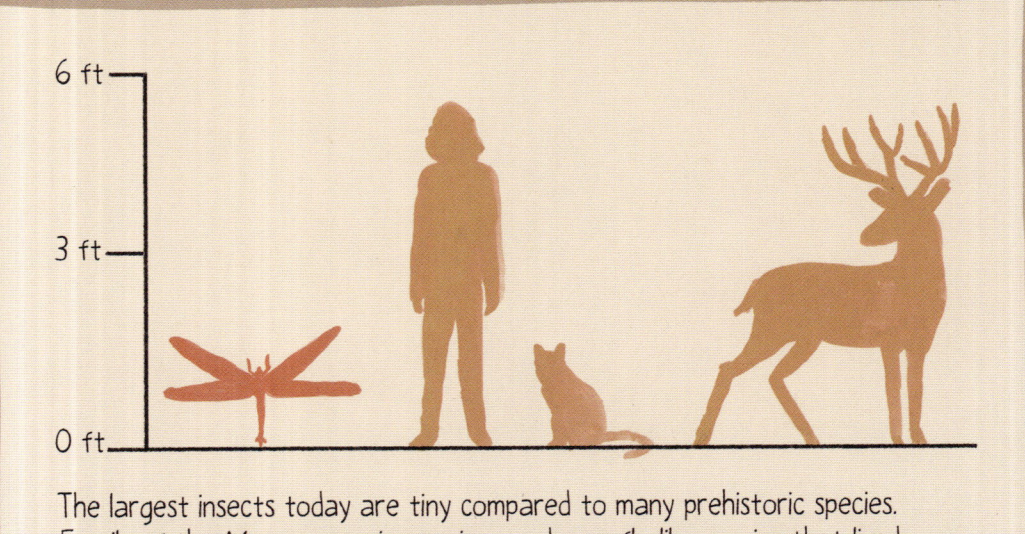

The largest insects today are tiny compared to many prehistoric species. Fossils of the *Meganeuropsis permiana*, a dragonfly-like species that lived 300 million years ago, suggest a wingspan up to 2.5 feet.

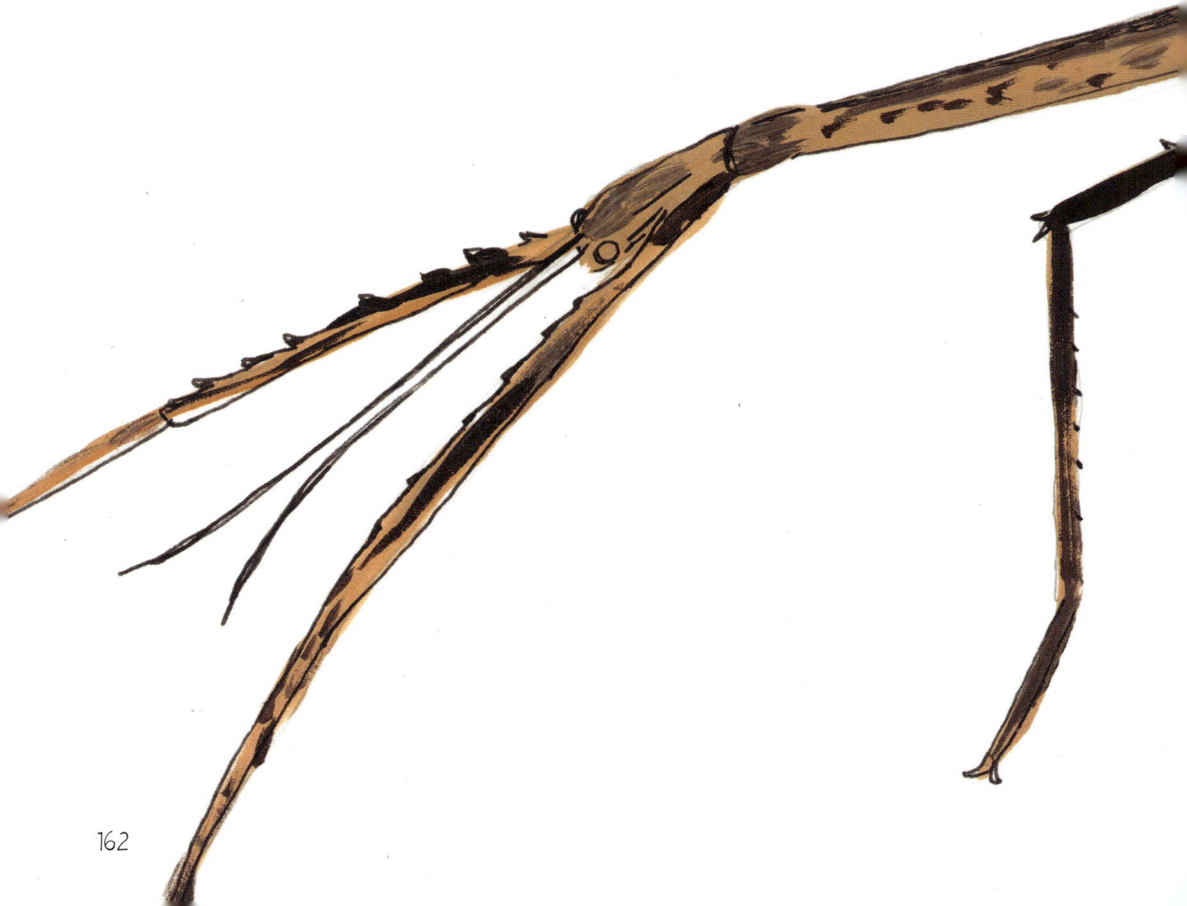

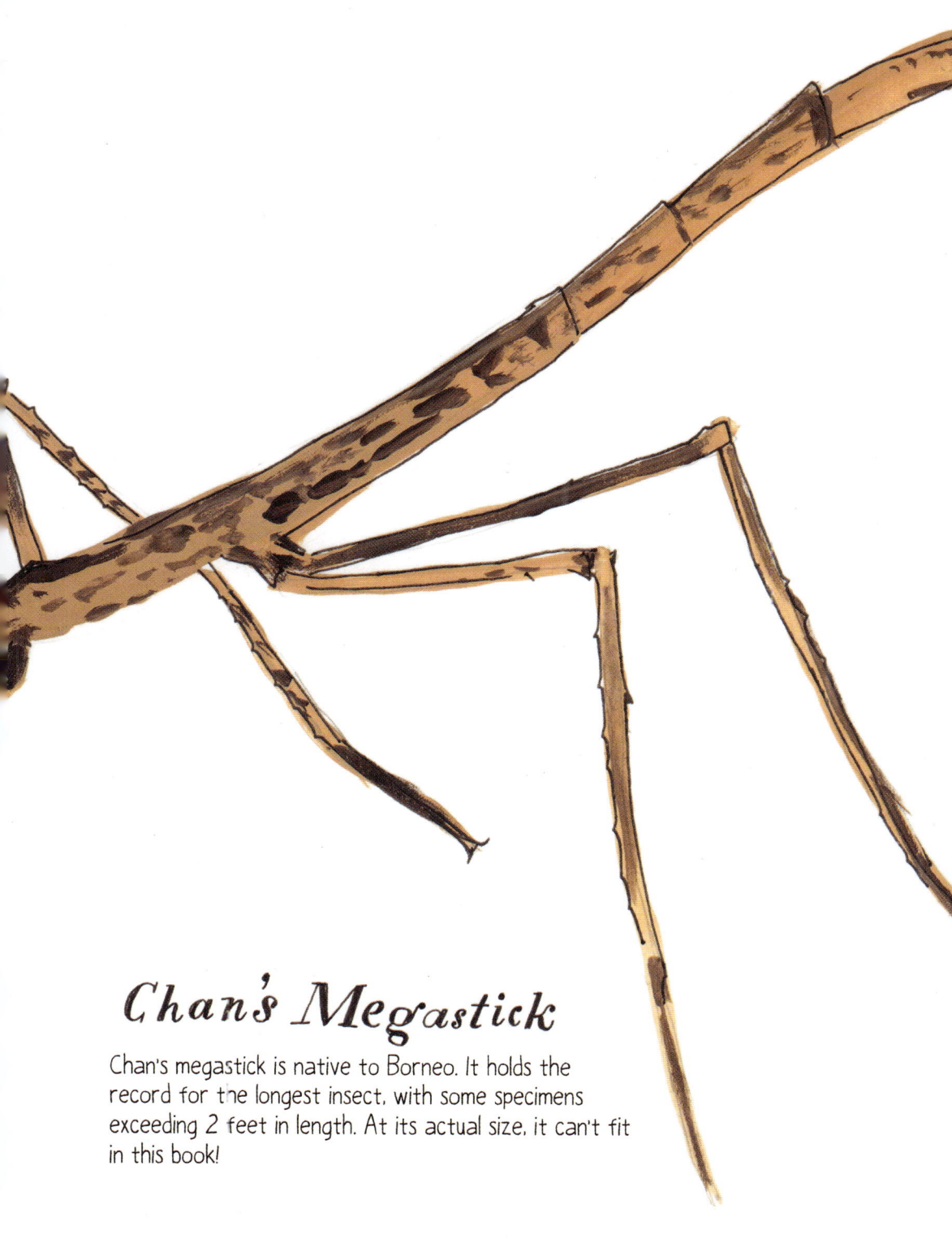

Chan's Megastick

Chan's megastick is native to Borneo. It holds the record for the longest insect, with some specimens exceeding 2 feet in length. At its actual size, it can't fit in this book!

HEAVIEST

The giant wētā is the heaviest insect in the world, reaching a weight of 2.5 ounces, about the same as a small hamster. This oversized grasshopper is found in New Zealand, where it has existed since the Mesozoic Era.

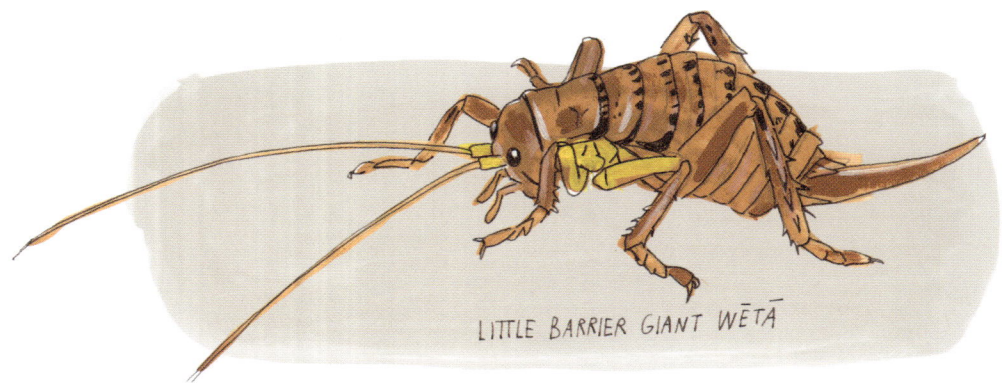

LITTLE BARRIER GIANT WĒTĀ

MOST RESILIENT

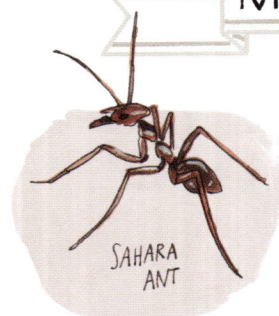

SAHARA ANT

Living in the heat of the Sahara, these ants can keep track of the sun's position and know the direct route back to their nest.

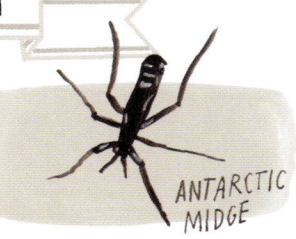

ANTARCTIC MIDGE

The Antarctic midge can survive freezing and lives year-round in Antarctica, where no other insect can.

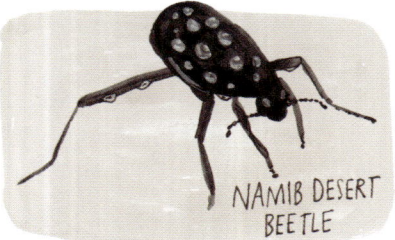

NAMIB DESERT BEETLE

A native to the Namib desert, where there is less than 0.55 inches of rain a year, this beetle survives by collecting water on its back from the fog.

CUTEST

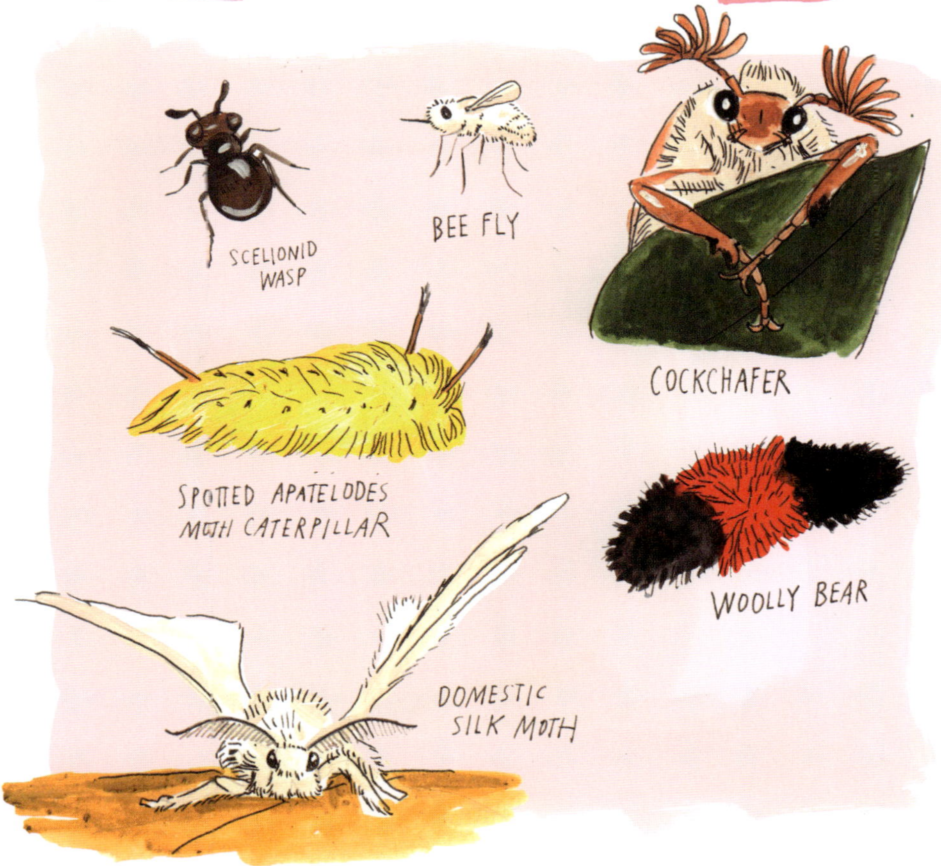

SCELIONID WASP

BEE FLY

COCKCHAFER

SPOTTED APATELODES MOTH CATERPILLAR

WOOLLY BEAR

DOMESTIC SILK MOTH

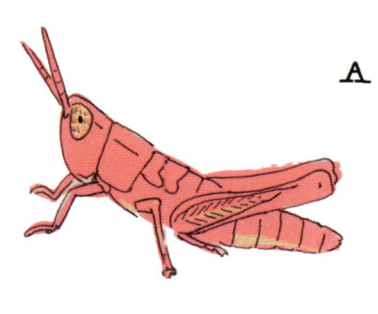

A PINK GRASSHOPPER?

Pink grasshoppers are extremely rare. Their unusual coloring, caused by a genetic mutation, makes it difficult for them to camouflage, leaving them more vulnerable to predators.

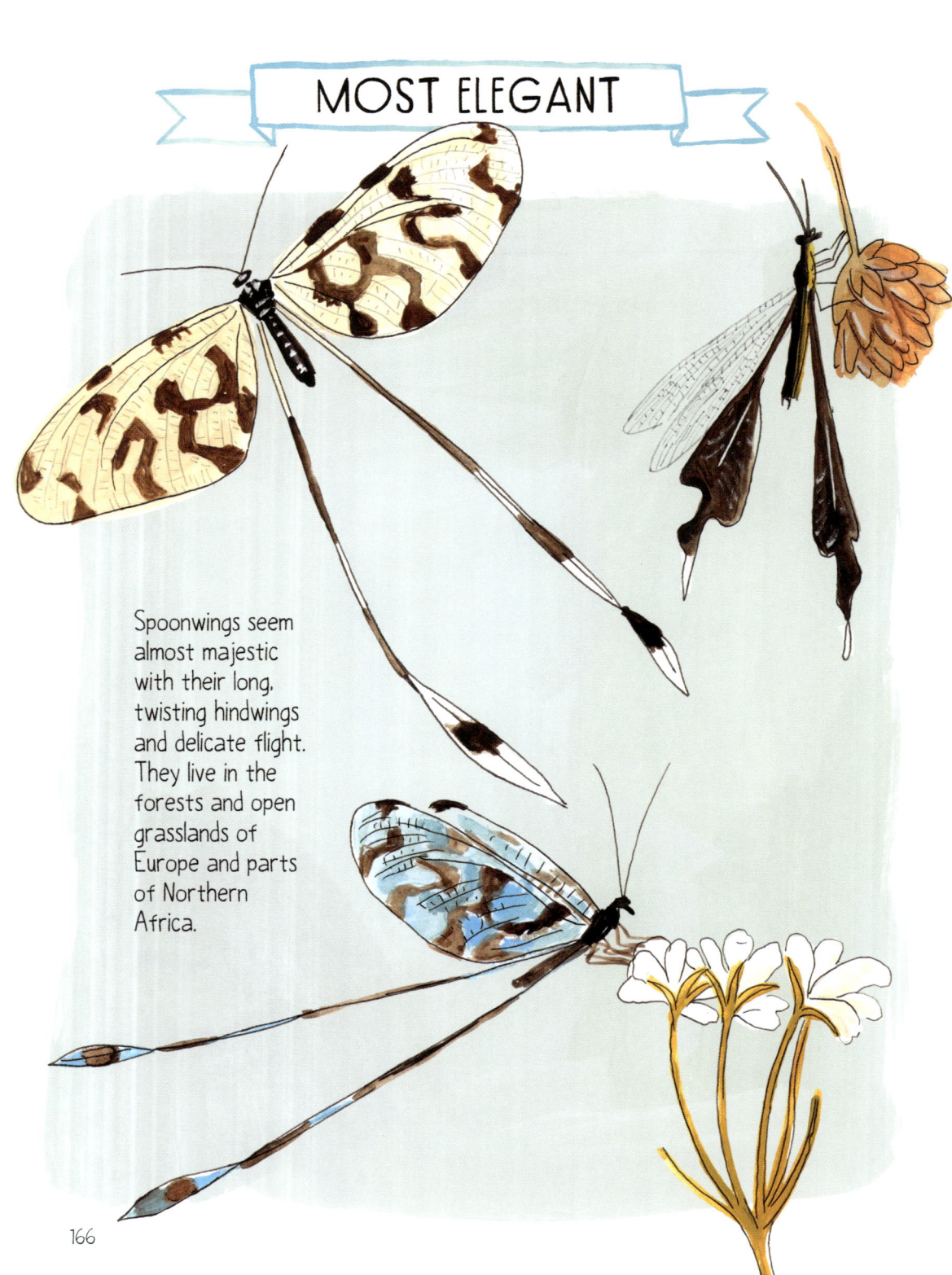

Spoonwings seem almost majestic with their long, twisting hindwings and delicate flight. They live in the forests and open grasslands of Europe and parts of Northern Africa.

BRIGHTEST

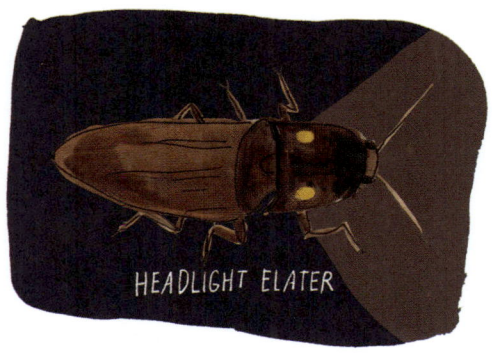

HEADLIGHT ELATER

This click beetle from the West Indies is thought to have the brightest bioluminescence of any insect, similar to that of a single LED flashlight.

SMALLEST

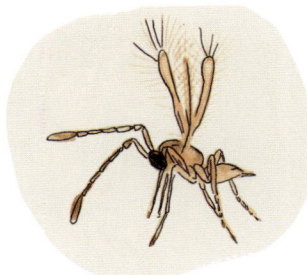

The fairy wasp is the smallest known insect at approximately 0.005 inches in length. Difficult to observe without a microscope, fairy wasps are parasitoids, laying their eggs within the eggs of other insects.

LONGEST-LIVING

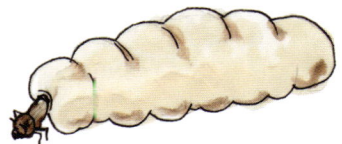

Termite queens can live between 30 and 50 years. Some scientists speculate they occasionally live to reach 100.

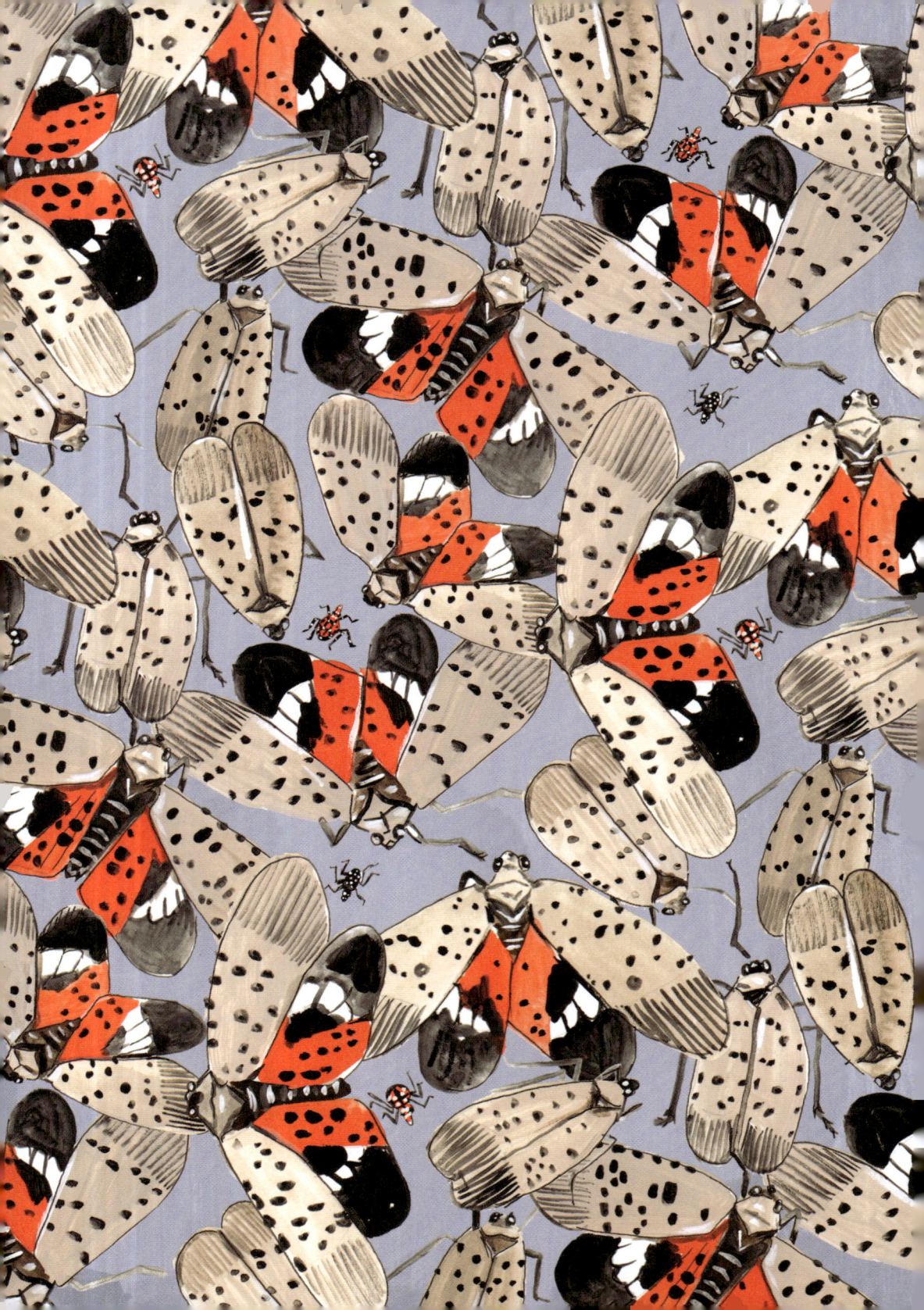

CHAPTER 7

They Are Among Us

SLEEP TIGHT, DON'T LET THE BEDBUGS BITE

We humans refer to insects as pests when they begin to threaten our health and comfort. One reason we build shelters is to keep our living spaces free from insects and other small creatures that may interfere with our sleep, our well-being, and perhaps our general sense of control. Nonetheless, insects often find a way into even the most protected environments.

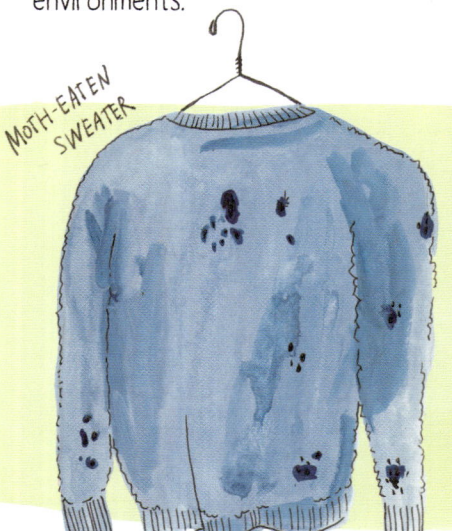

MOTH-EATEN SWEATER

CLOTHING MOTHS

Considered serious pests due to their hunger for clothing and natural fibers, these moths (actually, their larvae) are famously found in closets and dressers, chewing holes into garments.

BEDBUGS

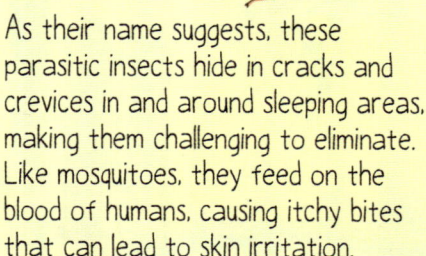

As their name suggests, these parasitic insects hide in cracks and crevices in and around sleeping areas, making them challenging to eliminate. Like mosquitoes, they feed on the blood of humans, causing itchy bites that can lead to skin irritation.

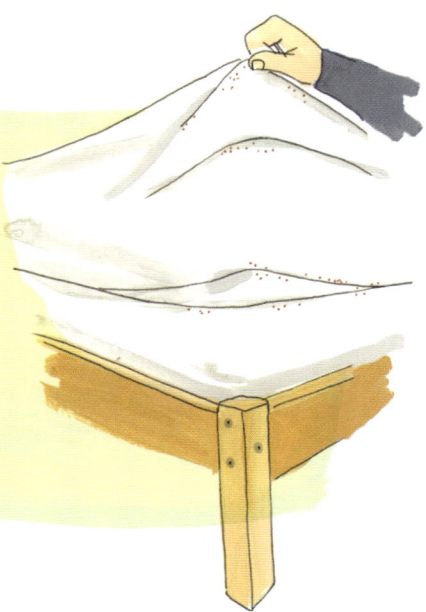

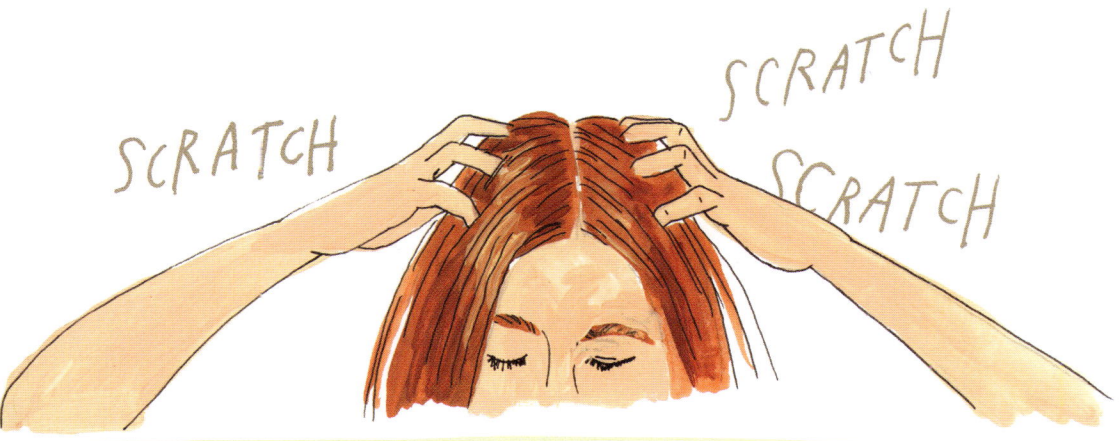

SCRATCH SCRATCH SCRATCH

LICE

Lice are small wingless insects that live on the hair and skin of humans or animals, feeding on blood. Infestations can lead to itching, irritation, and potential skin infections. Lice can spread easily from person to person through close contact.

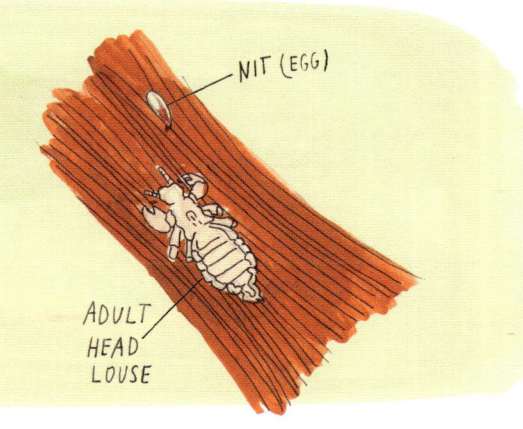

NIT (EGG)

ADULT HEAD LOUSE

FLEAS

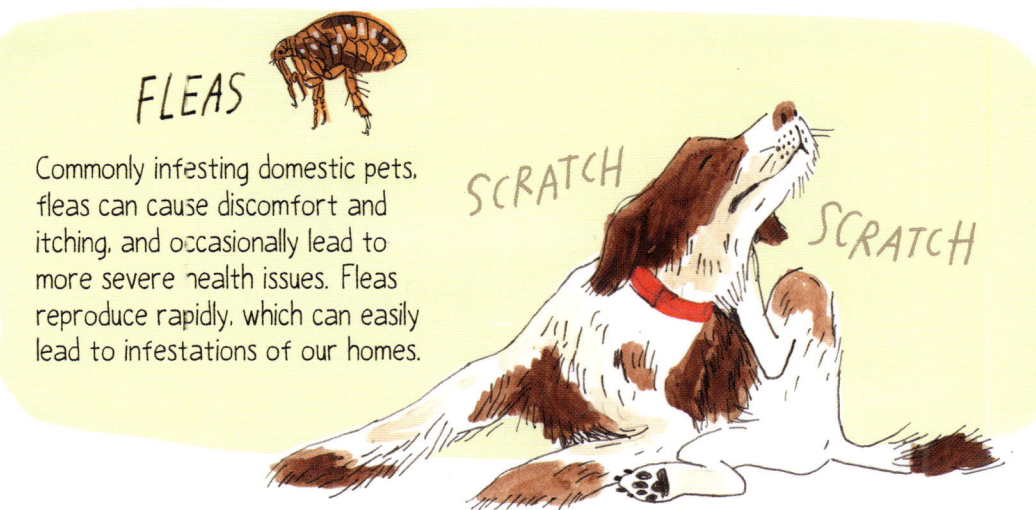

Commonly infesting domestic pets, fleas can cause discomfort and itching, and occasionally lead to more severe health issues. Fleas reproduce rapidly, which can easily lead to infestations of our homes.

SCRATCH SCRATCH

ANATOMY OF A MOSQUITO

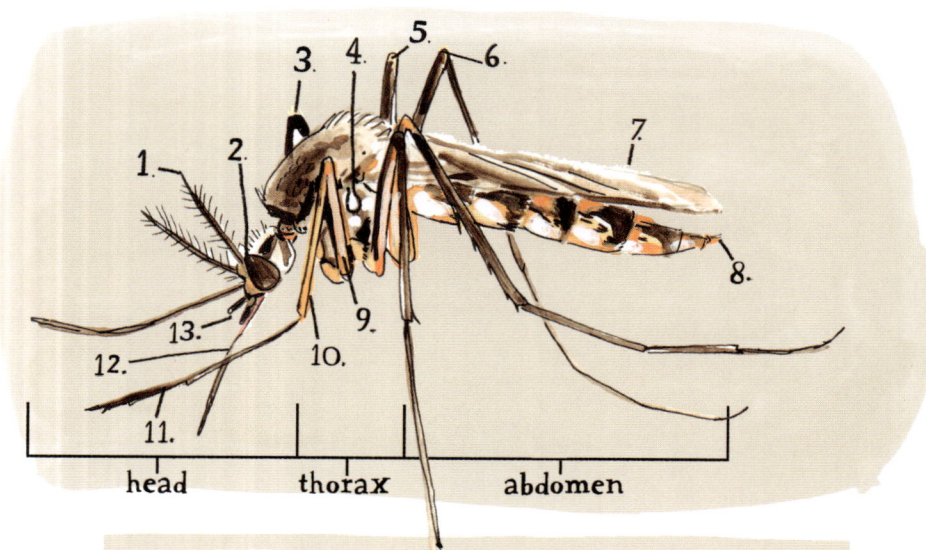

head   thorax   abdomen

1. antenna
2. compound eye
3. foreleg
4. haltere
5. midleg
6. hind leg
7. wing

8. cerci
9. femur
10. tibia
11. tarsus
12. proboscis
13. palp

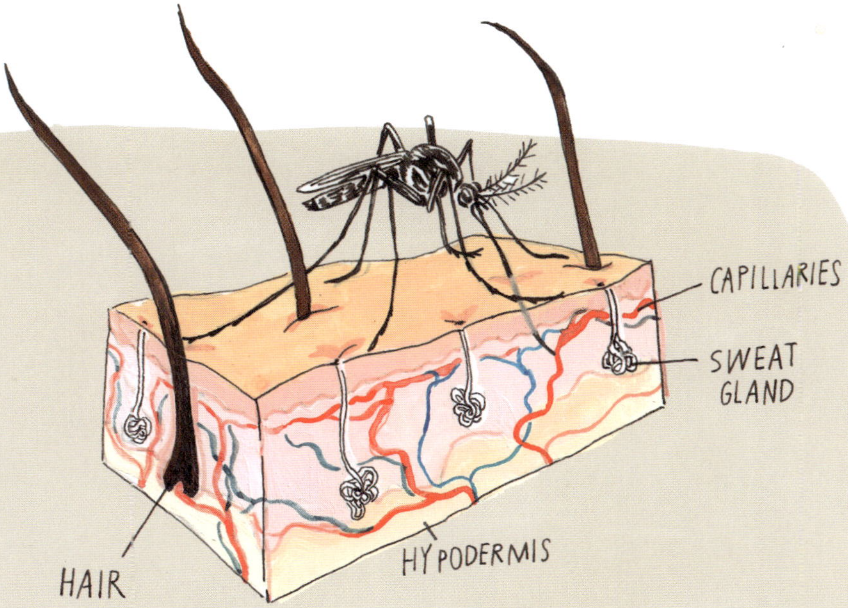

CAPILLARIES

SWEAT
GLAND

HAIR

HYPODERMIS

Mosquitoes use their needle-like proboscis to pierce the skin of their victims and suck up blood. While feeding, they also inject saliva into the skin, which can cause significant discomfort, including itching and allergic reactions.

Mosquitoes cause by far the most human deaths of any animal, with 750,000 to 1 million fatalities a year from diseases, especially malaria.

A LIST OF DISEASES SPREAD BY BITING INSECTS:

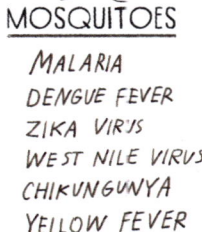

MOSQUITOES

MALARIA
DENGUE FEVER
ZIKA VIRUS
WEST NILE VIRUS
CHIKUNGUNYA
YELLOW FEVER

FLEAS

MURINE TYPHUS
BUBONIC PLAGUE

LICE
TRENCH FEVER

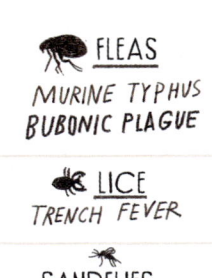

SANDFLIES
LEISHMANIASIS

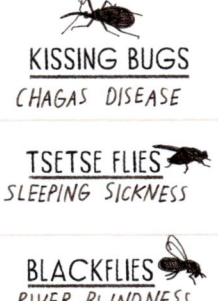

KISSING BUGS
CHAGAS DISEASE

TSETSE FLIES
SLEEPING SICKNESS

BLACKFLIES
RIVER BLINDNESS

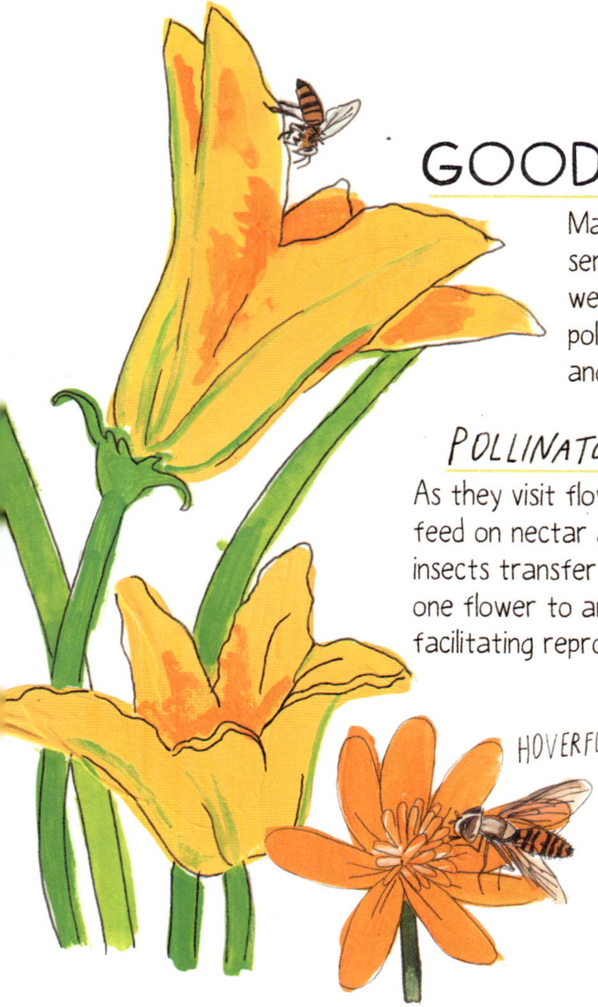

GOOD FOR THE GARDEN

Many species of insects perform valuable services to ecosystems and human well-being by playing essential roles in pollination, pest control, decomposition, and nutrient cycling.

POLLINATORS

As they visit flowers to feed on nectar and pollen, insects transfer pollen from one flower to another, facilitating reproduction.

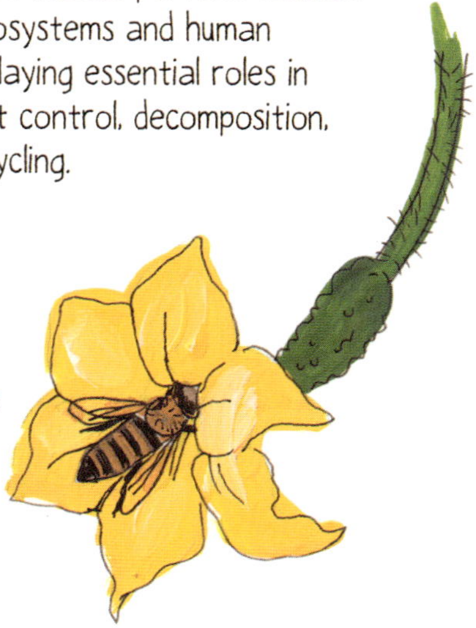

HOVERFLY

There are approximately 20,000 species of bees in the world, at least 3,600 of which are native to North America. More than 90 percent of those are solitary bees, which typically don't sting.

DECOMPOSERS

In addition to preying on live insects, some beetles feed on dead plants and other decomposing organic matter, which helps to recycle nutrients back into the ecosystem. Some beetles also tunnel through the soil, which can help to improve soil aeration.

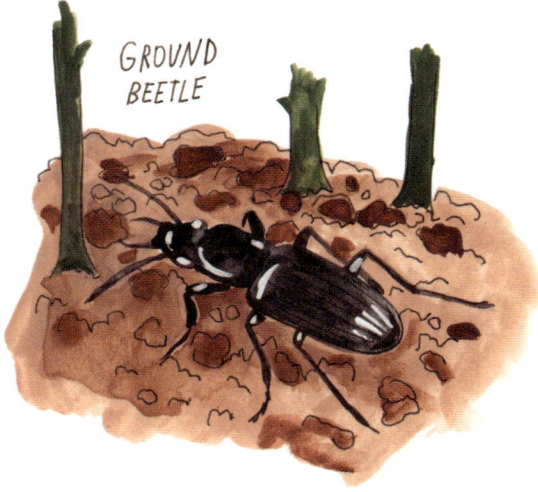

GROUND BEETLE

PREDATORS

Predatory insects provide pest control in gardens and agricultural fields against crop-eating insects like aphids and mites, reducing the need for chemical pesticides and promoting sustainable farming practices.

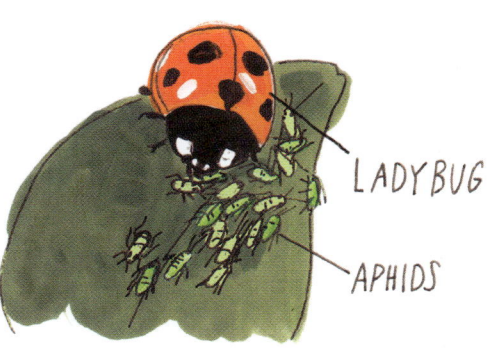

A ladybug will eat 50 aphids a day and thousands over its lifetime.

LADYBUG

APHIDS

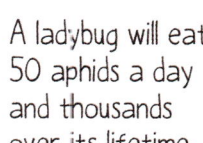

LACEWING LARVA

The larva of a lacewing bug, often called an "aphid lion" uses its large, curved mandibles to pierce and extract juices from aphids. It devours about 200 aphids a week.

APHIDS

Spined soldier bugs are voracious predators that use their beaks to jab and inject digestive enzymes that paralyze their prey. Then they suck out their insides.

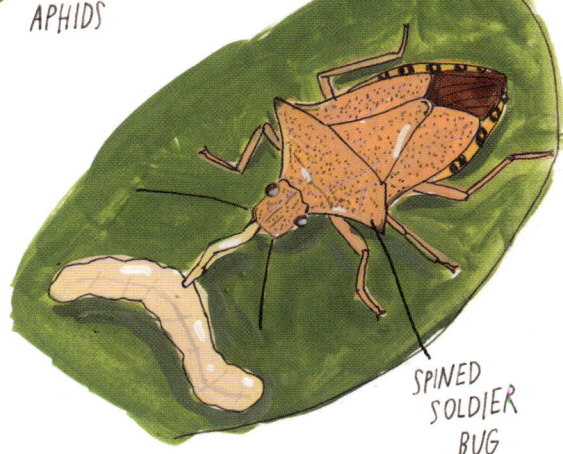

SPINED SOLDIER BUG

BAD FOR THE GARDEN

Referring to insects as pests depends on the context and circumstances. While certain species may cause significant damage to a farmer's crops, they may simultaneously play a beneficial role in another ecosystem.

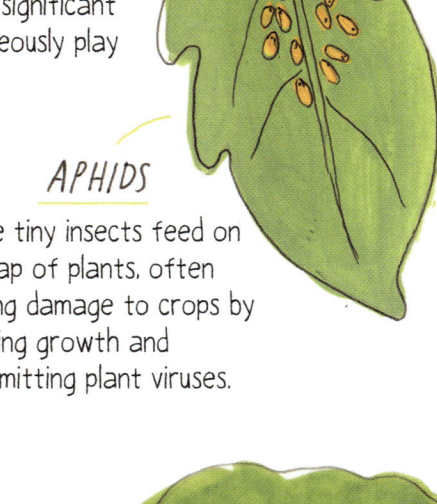

APHIDS

These tiny insects feed on the sap of plants, often causing damage to crops by stunting growth and transmitting plant viruses.

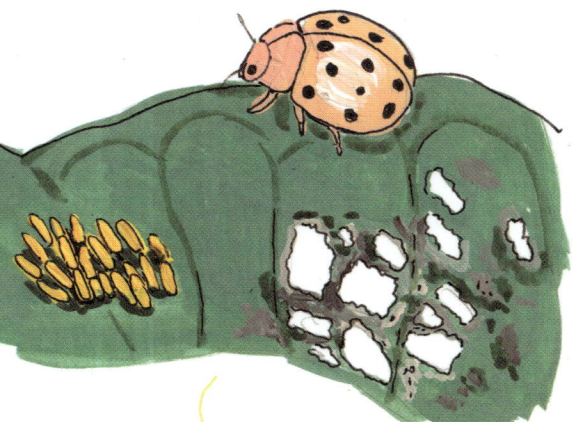

MEXICAN BEAN BEETLE

These are the only lady beetles that are considered pests to farmers. They feed on bean and other legume plants.

CORN EARWORM

The larvae of this moth feed on the leaves, tassels, and mostly the inside of ears of corn.

CABBAGE WORM

This larval pest not only consumes cabbage, it can damage a variety of crops, including broccoli and Brussels sprouts, as well as certain fruits.

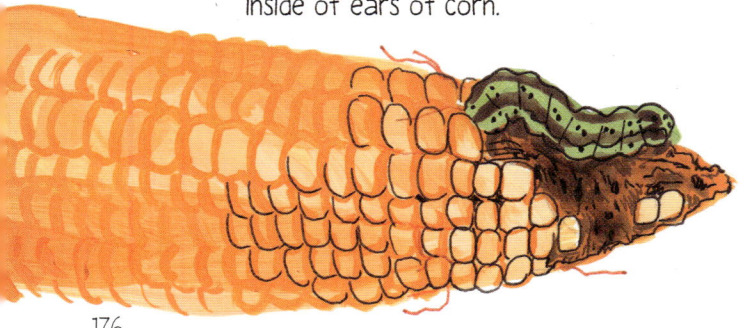

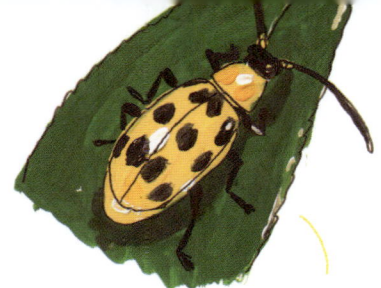

SPOTTED CUCUMBER BEETLE

Whether striped or spotted, these beetles can damage the foliage and fruit of cucumber plants and transmit bacteria that cause the plants to die.

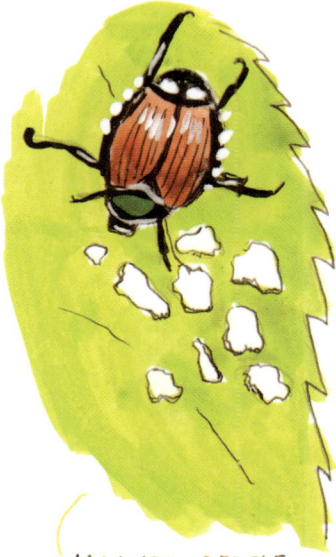

JAPANESE BEETLE

The grubs of Japanese beetles damage lawns and pastures, while the adults prefer the flowers and fruits of both ornamental and agricultural plants.

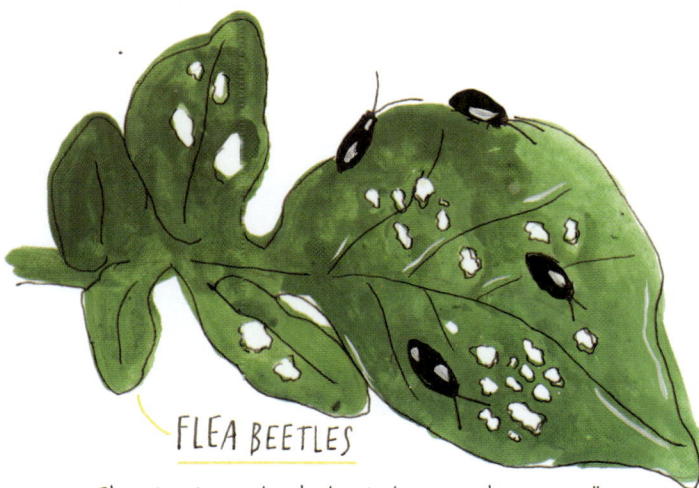

FLEA BEETLES

Chewing irregular holes in leaves, these small beetles damage a range of crops from radishes to broccoli to spinach.

COLORADO POTATO BEETLE

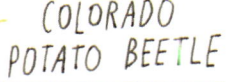

These beetles significantly damage potato, eggplant, and pepper plants. Luckily stink bugs and lady beetles will prey on their eggs.

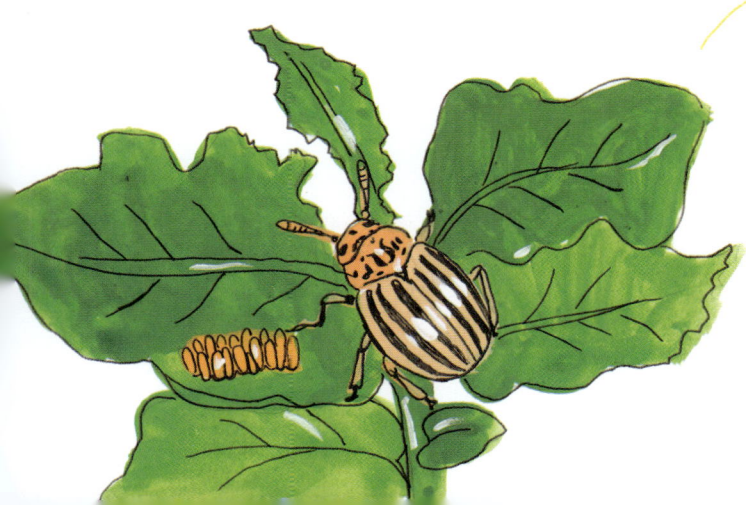

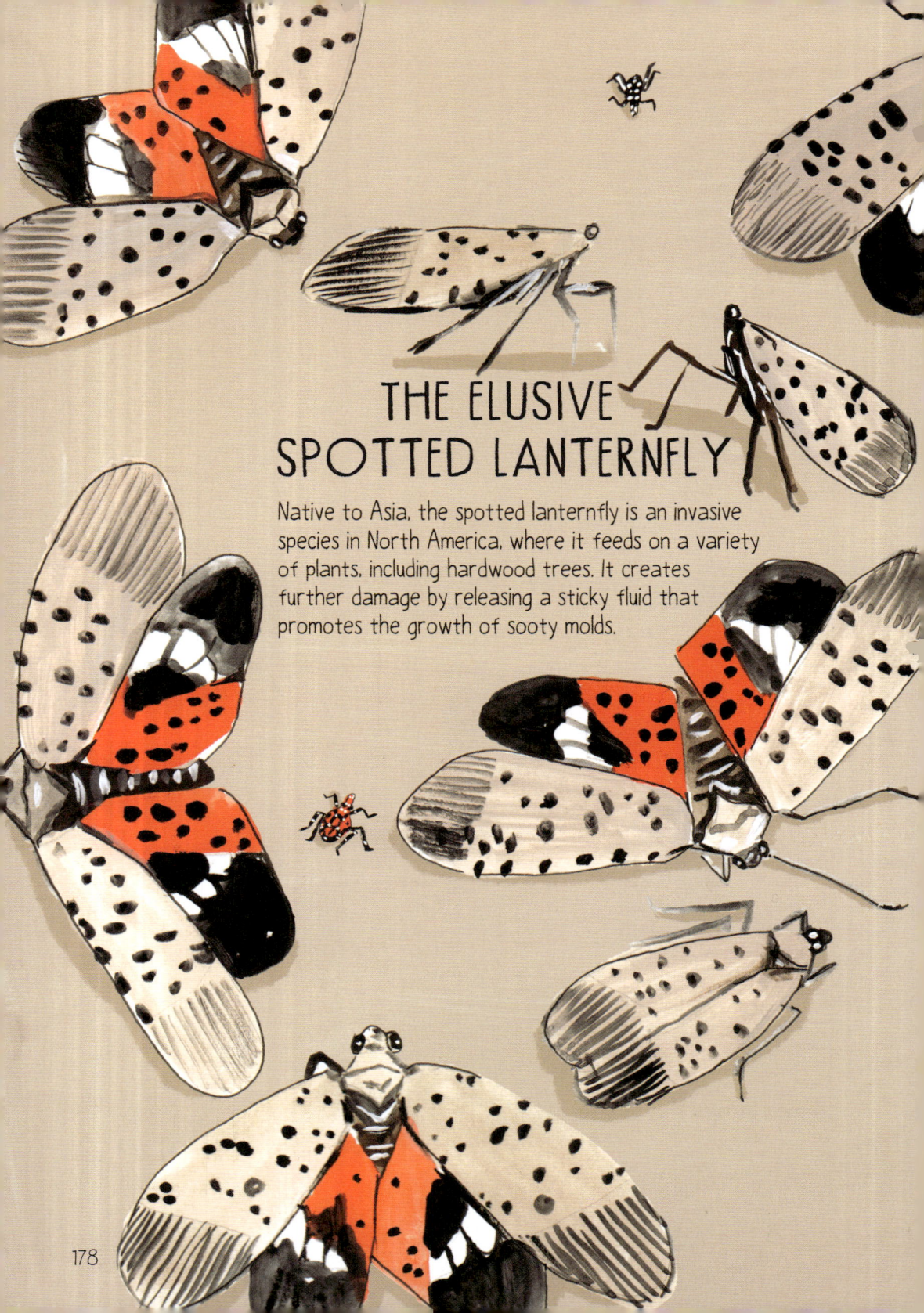

THE ELUSIVE SPOTTED LANTERNFLY

Native to Asia, the spotted lanternfly is an invasive species in North America, where it feeds on a variety of plants, including hardwood trees. It creates further damage by releasing a sticky fluid that promotes the growth of sooty molds.

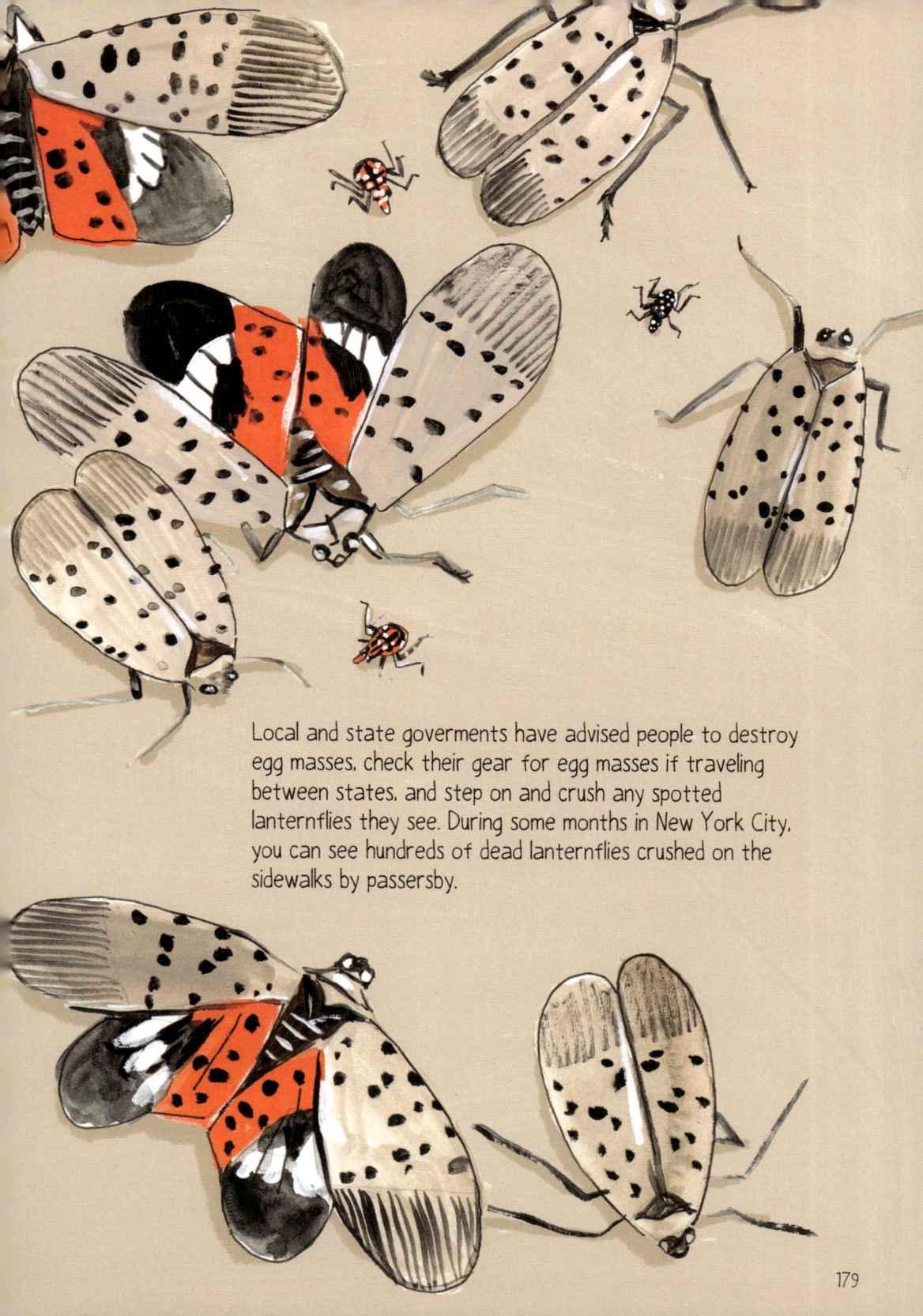

Local and state goverments have advised people to destroy egg masses, check their gear for egg masses if traveling between states, and step on and crush any spotted lanternflies they see. During some months in New York City, you can see hundreds of dead lanternflies crushed on the sidewalks by passersby.

CHIRPING IN CHINA

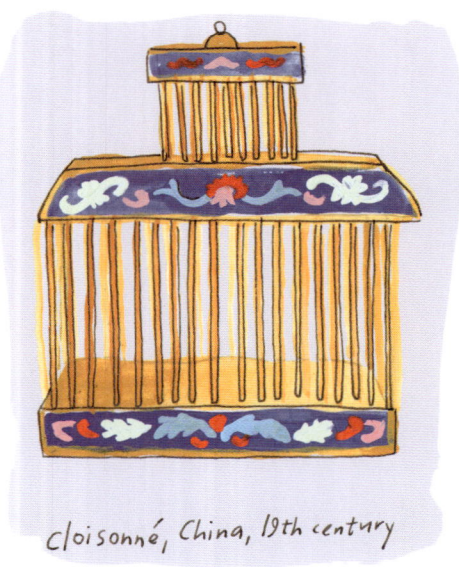

cloisonné, China, 19th century

People in China have been keeping crickets as pets from at least as long ago as the Tang Dynasty (618-906 CE). Crickets were viewed as symbols of good fortune and happiness. They were kept in beautifully decorated cages made of fine materials.

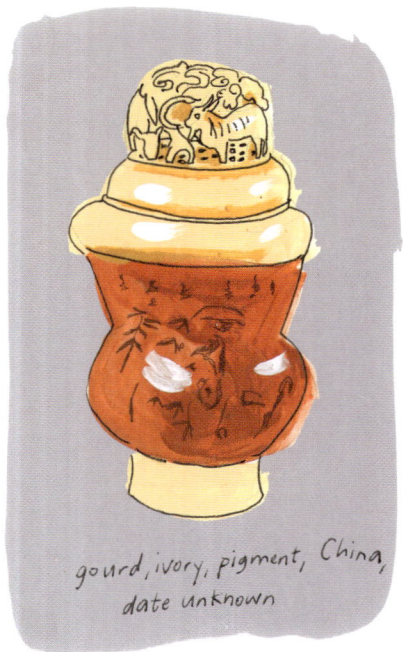

gourd, ivory, pigment, China, date unknown

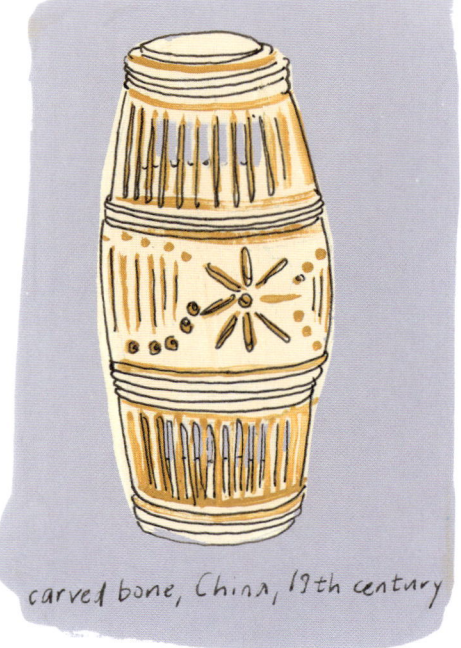

carved bone, China, 19th century

INSECT EATERS

Protein-rich insects are eaten by all kinds of animals. An insectivore is an animal that feeds on insects as its primary nutrition.

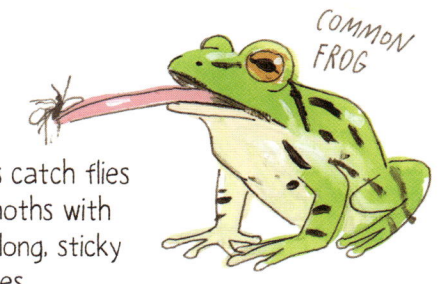

COMMON FROG

Frogs catch flies and moths with their long, sticky tongues.

LITTLE BROWN BAT

Small bats might eat 4 to 8 grams of moths, mosquitoes, flies, and beetles each night.

GIANT ANTEATER

Anteaters tear open ant and termite nests with their claws, then use their tongue to lap up about 30,000 insects a day.

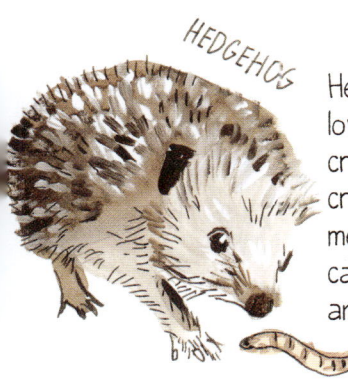

HEDGEHOG

Hedgehogs love little creepy crawlers like mealworms, caterpillars, and crickets.

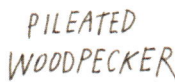

PILEATED WOODPECKER

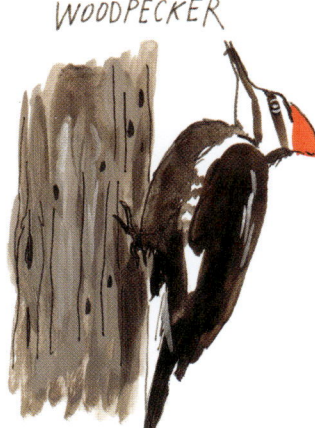

Some woodpeckers create holes in trees to get at the various insects and their larvae that live under the bark.

181

ON THE MENU

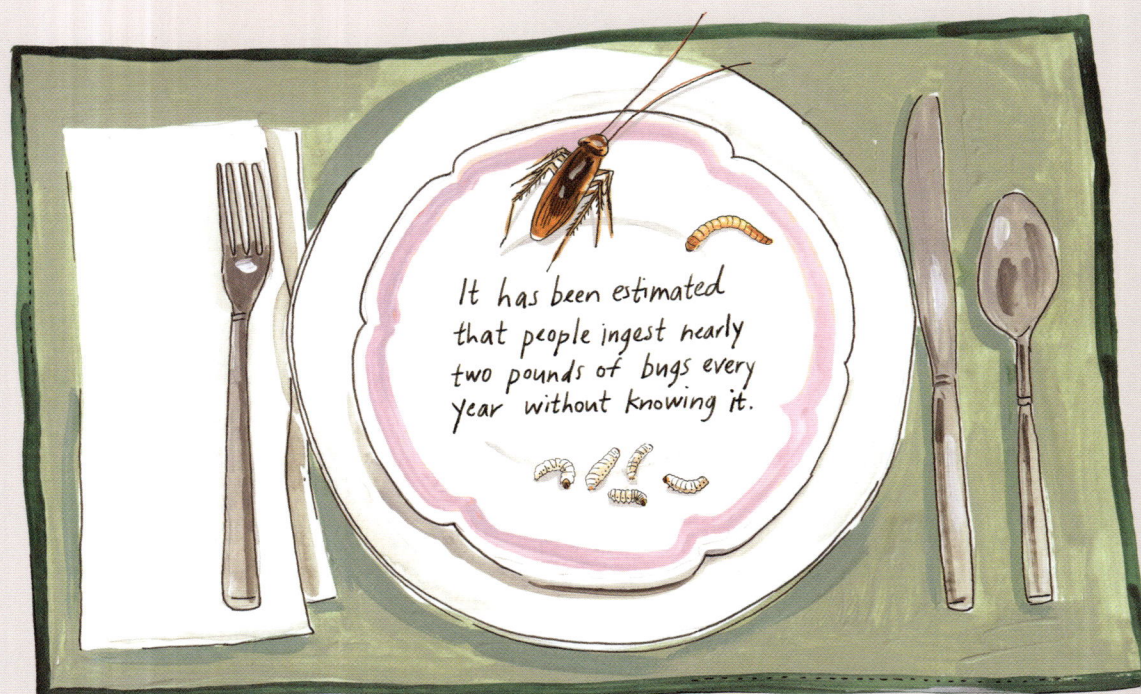

It has been estimated that people ingest nearly two pounds of bugs every year without knowing it.

The FDA accepts the fact that everyday foods such as baked goods and coffee may contain bits of insects such as mealworms, maggots, and roaches.

Many bugs are perfectly safe to eat and can provide substantial nutrition. Europe and North America are among the only places where eating insects is not common practice.

The practice of eating insects is known as entomophagy.

Some types of insects that are regularly consumed by humans around the world include ants, beetles, caterpillars, centipedes, cicadas, cockroaches, crickets, dragonflies, grasshoppers, June bugs, locusts, mealworms, midge flies, pill bugs, stink bugs, termites, walking sticks, and wasps.

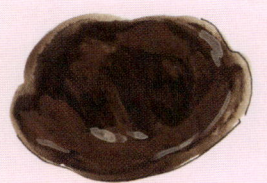

KUNGA CAKE

In parts of Africa, tiny swarming midge flies are captured using large pots and nets. They are then smashed into a paste to form dense patties or "kunga cakes" that are grilled.

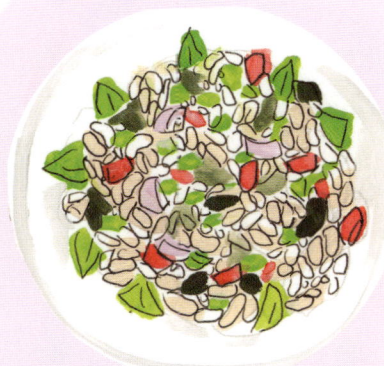

KOI KHAI MOT DANG
(SPICY RAW ANT EGGS SALAD)

Both the pupae and eggs of the weaver ant are consumed in several Southeast Asian countries, including Thailand and Laos. The eggs are a source of protein and are typically enjoyed in soups and salads, adding a unique sour flavor and popping texture.

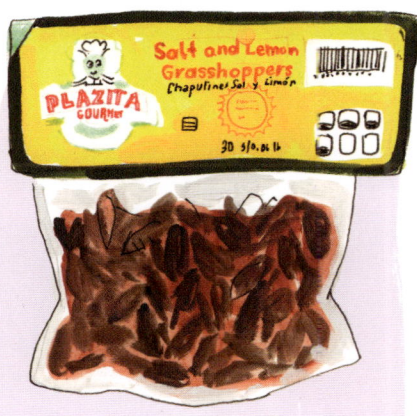

CHAPULINES

Chapulines (roasted grasshoppers) are common fare in Oaxaca, Mexico. They're often eaten as snacks or crushed and sprinkled on top of dishes to enhance the flavor.

WE ATE INSECTS!

While working on this chapter, Michael suggested we try eating insects as part of our research. I'm a vegetarian, so it wasn't something I wanted to do, but I tried one tiny bite of a chocolate-covered ant. Thankfully, it just tasted like chocolate.

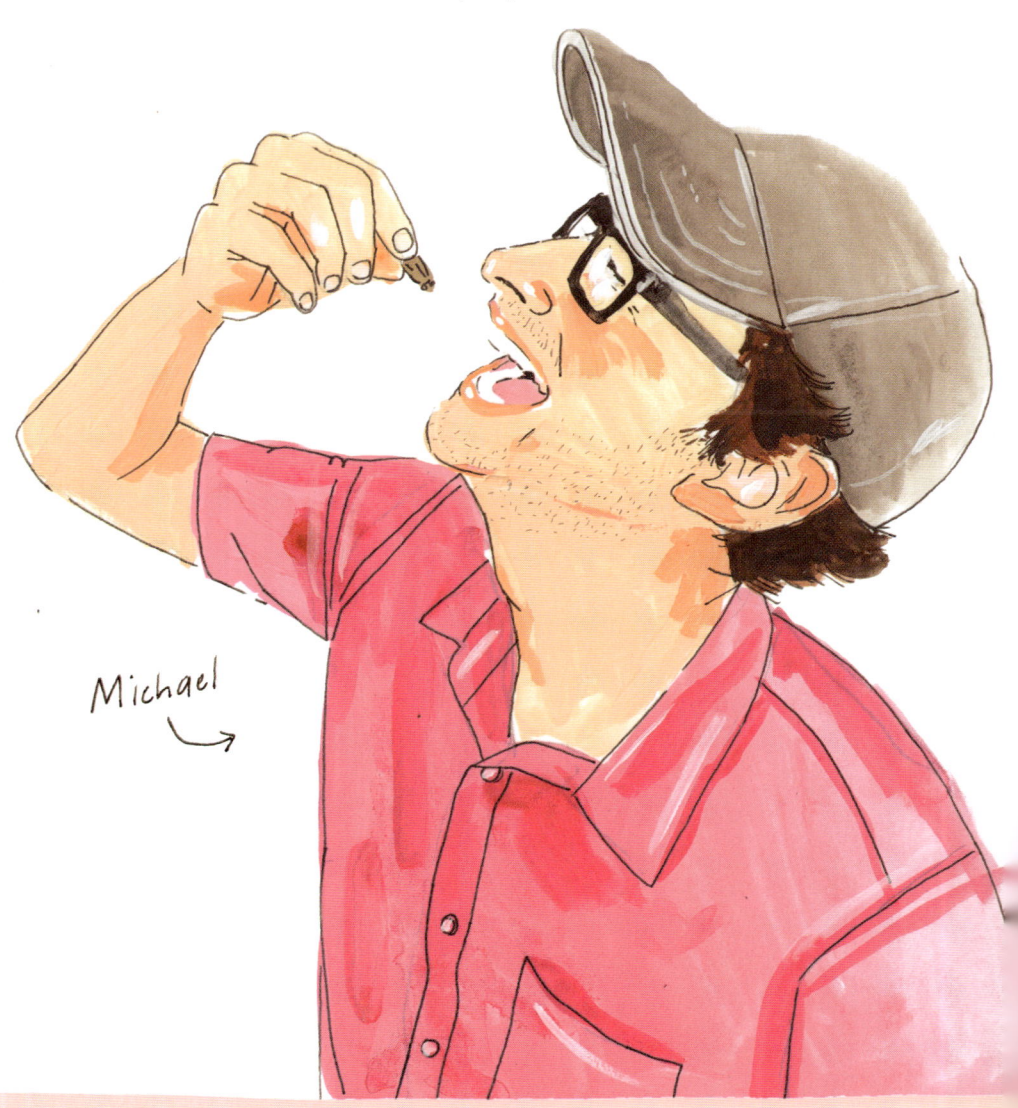

Michael →

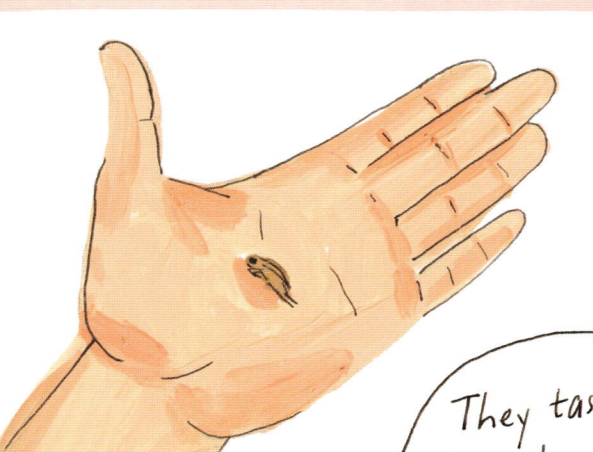

Michael chomped crickets and mealworms and chocolate-covered ants. Here are some things he said while eating the dried insects:

They taste like chocolate-covered Rice Krispies.

They're really good! So crunchy.

I prefer the crickets. These don't have as much flavor.

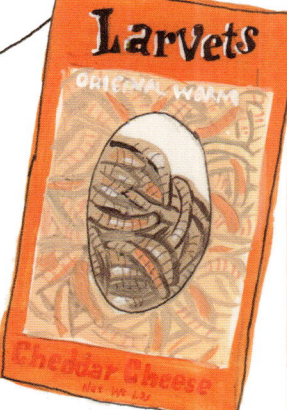

We found these packaged insect snacks at a candy shop in our Brooklyn neighborhood. The shop owner said the insects are popular and there are a lot of repeat customers who come back for them again and again.

It's safe to say we won't be those repeat customers.

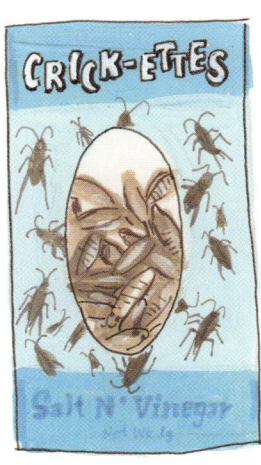

Everyday Products Made with Insects

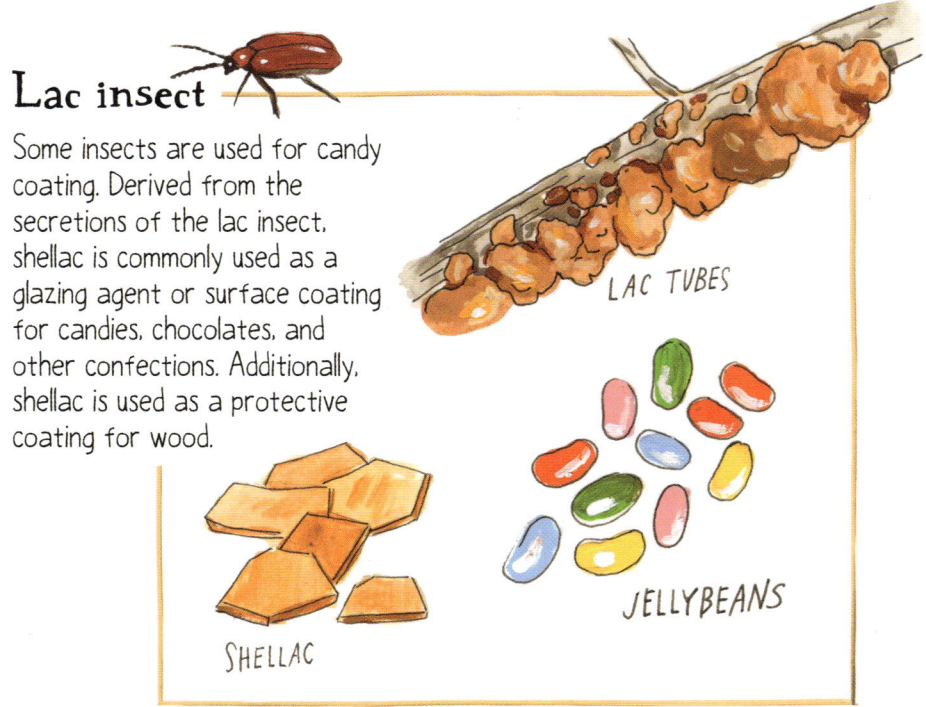

Lac insect

Some insects are used for candy coating. Derived from the secretions of the lac insect, shellac is commonly used as a glazing agent or surface coating for candies, chocolates, and other confections. Additionally, shellac is used as a protective coating for wood.

LAC TUBES

JELLYBEANS

SHELLAC

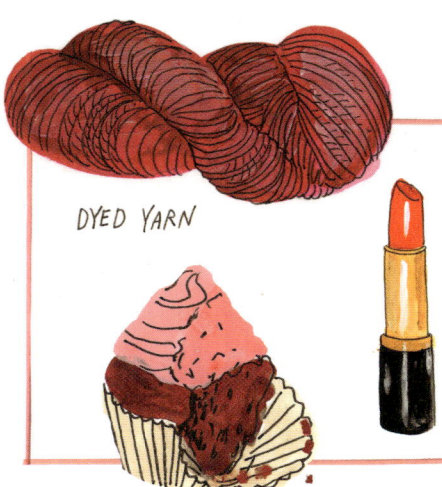

DYED YARN

Cochineal bug

The cochineal insect has been used for centuries to create a vivid red pigment called carmine. The insects are harvested, dried, and then crushed to make powder. This powdered pigment is used to dye cosmetics, textiles, and food coloring.

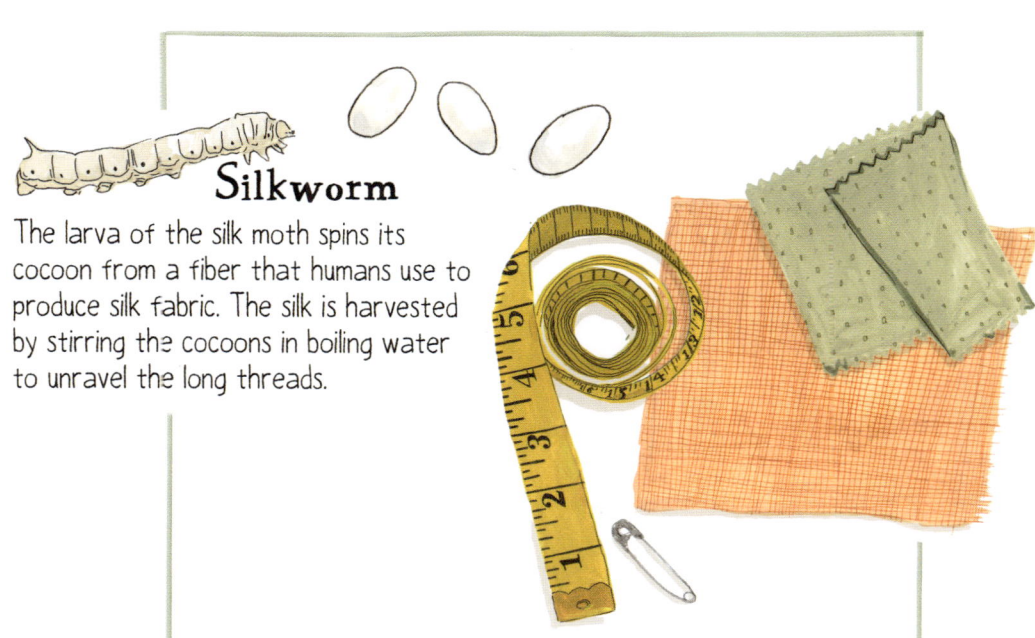

Silkworm

The larva of the silk moth spins its cocoon from a fiber that humans use to produce silk fabric. The silk is harvested by stirring the cocoons in boiling water to unravel the long threads.

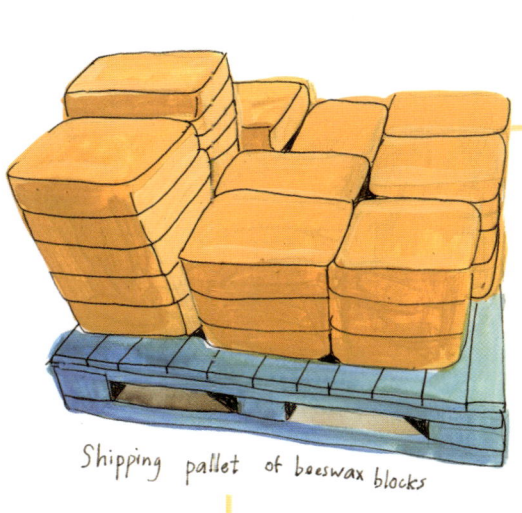

Shipping pallet of beeswax blocks

Honey bee

Over 60,000 tons of beeswax is harvested around the world every year from worker honey bees. India produces the largest share of that—about 23,000 tons. Beeswax forms the base of many common consumer products, from cosmetics and skincare items to candles to polishes and waxes for wood. It's used to make natural crayons and other art supplies.

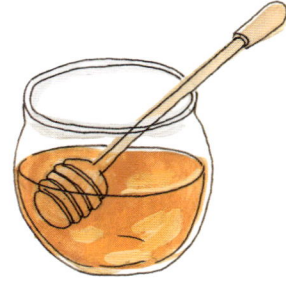

BEEKEEPING

Humans have practiced beekeeping, or apiculture, for at least 10,000 years. Depictions of beekeeping are found in Egyptian art. Bees are kept in artificial hives for the commercial production of honey and wax and the pollination of crops. Beekeeping is also a popular hobby.

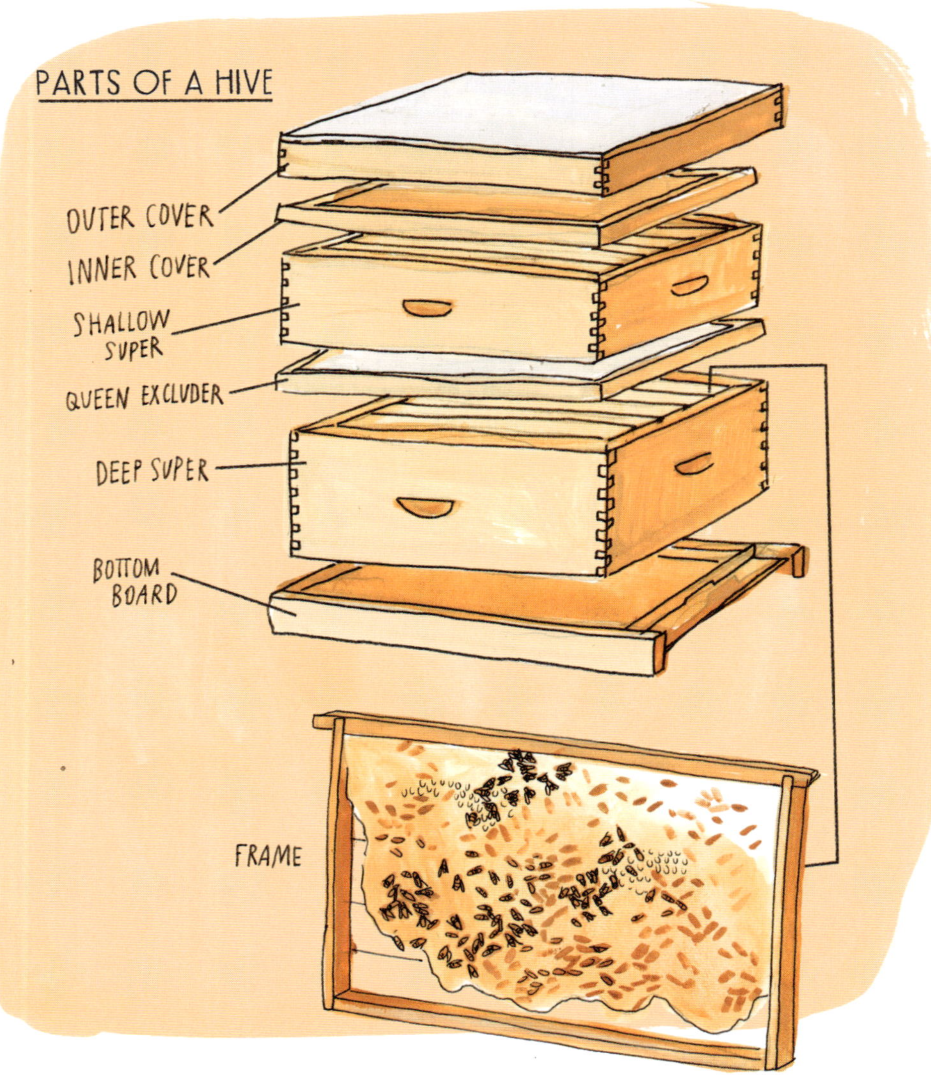

PARTS OF A HIVE

OUTER COVER

INNER COVER

SHALLOW SUPER

QUEEN EXCLUDER

DEEP SUPER

BOTTOM BOARD

FRAME

Beekeepers protect themselves from stings by wearing gloves, a full body suit, and a hat with a veil.

A variety of tools are used in beekeeping, most notably a smoker. A smoker has a firebox where pine needles, cardboard, or another natural material is smoldered to make smoke. The smoke calms the bees, allowing the beekeeper to work safely.

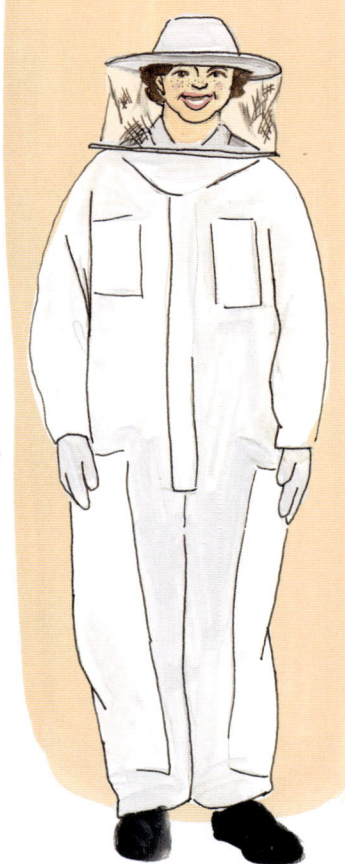

BEEKEEPER

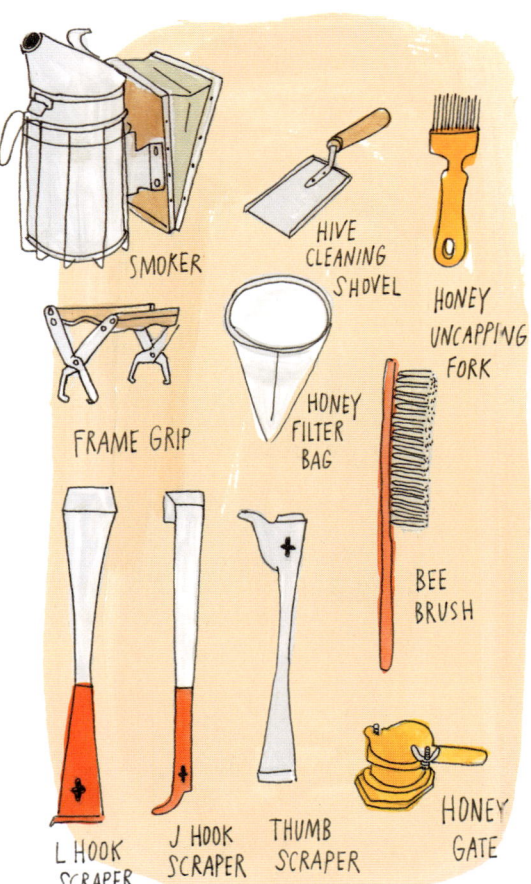

SMOKER

HIVE CLEANING SHOVEL

HONEY UNCAPPING FORK

FRAME GRIP

HONEY FILTER BAG

BEE BRUSH

L HOOK SCRAPER

J HOOK SCRAPER

THUMB SCRAPER

HONEY GATE

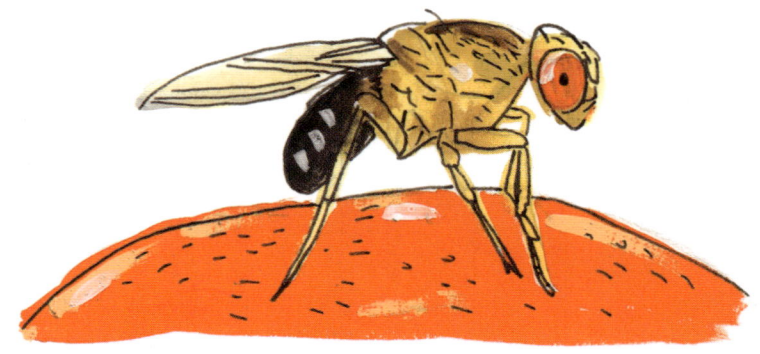

TINY FLIES, BIG DISCOVERIES

Fruit flies are an invaluable organism for scientific research. Sharing many of the same genes that cause diseases in humans, fruit flies help scientists understand how diseases develop and spread. They have a life cycle of only about 10 days, so scientists are able to observe multiple generations of flies.

Thomas Hunt Morgan, a pioneering geneticist from the early 20th century, used fruit flies in experiments to help him discover how traits are passed down through chromosomes (like eye color, for example). His findings earned him a Nobel Prize in 1933.

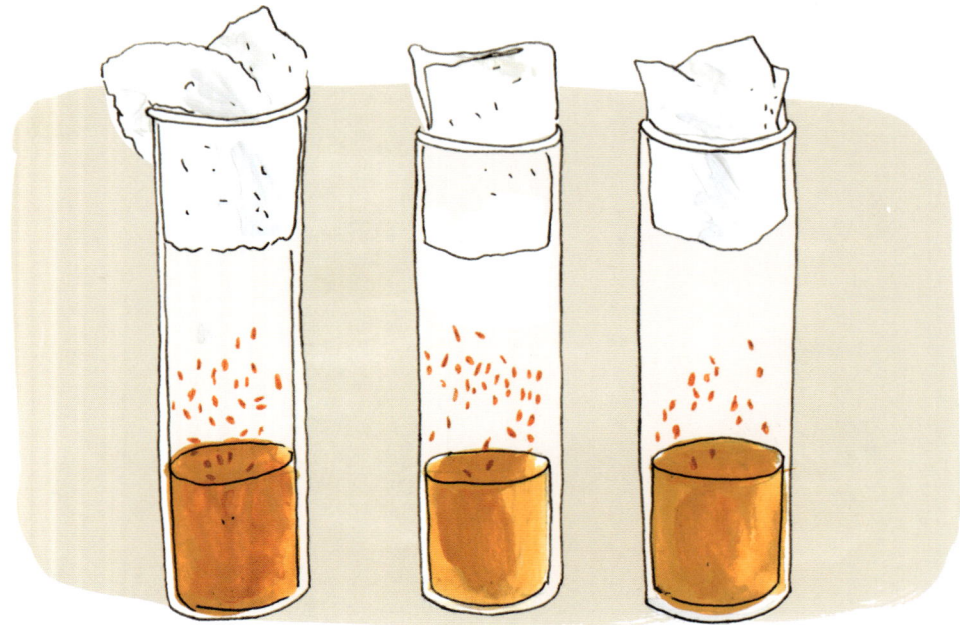

FLIES

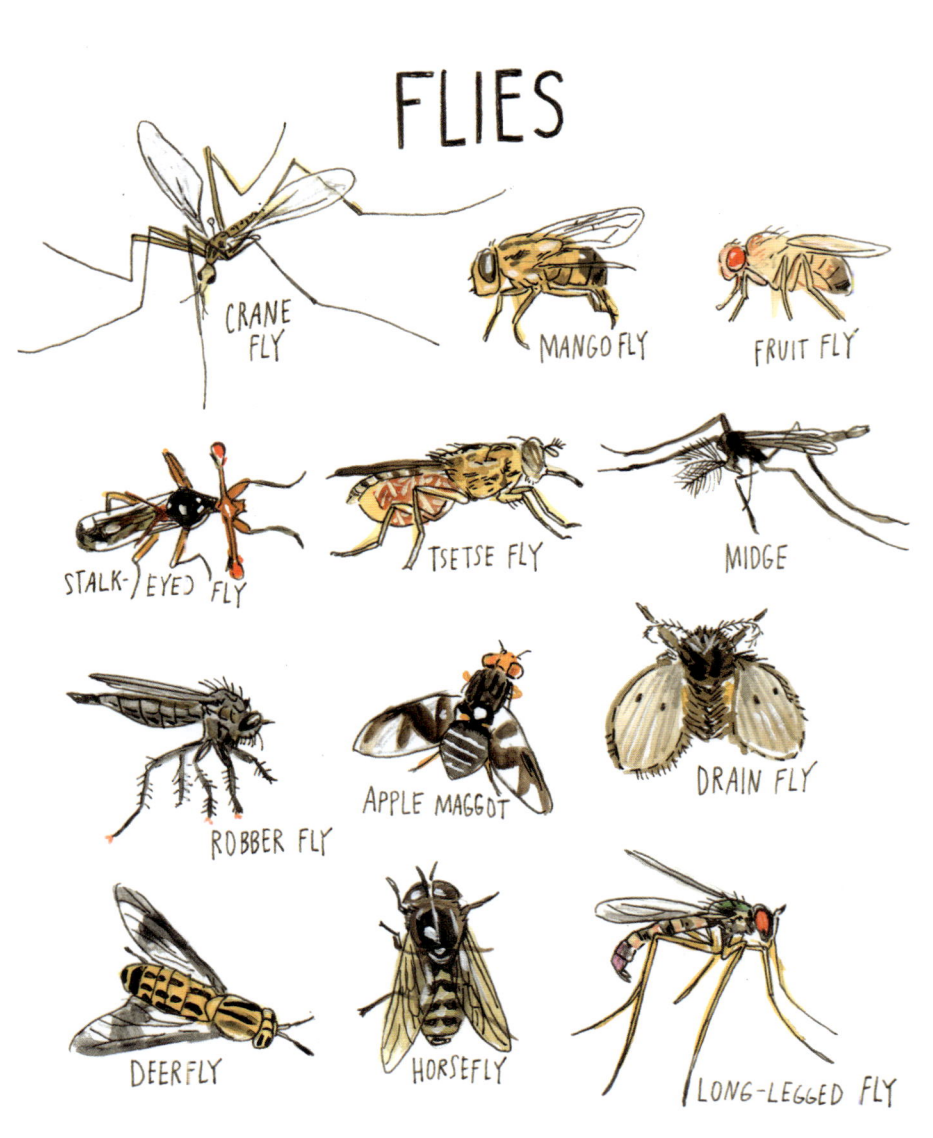

CRANE FLY

MANGO FLY

FRUIT FLY

STALK-EYED FLY

TSETSE FLY

MIDGE

ROBBER FLY

APPLE MAGGOT

DRAIN FLY

DEERFLY

HORSEFLY

LONG-LEGGED FLY

A fly's reaction time is about 12 times faster than a human's, which is why it's so difficult to swat one successfully. The fastest reflex belongs to the long-legged fly, which can respond in just 5 milliseconds.

MEET DR. SUNDAY EKESI

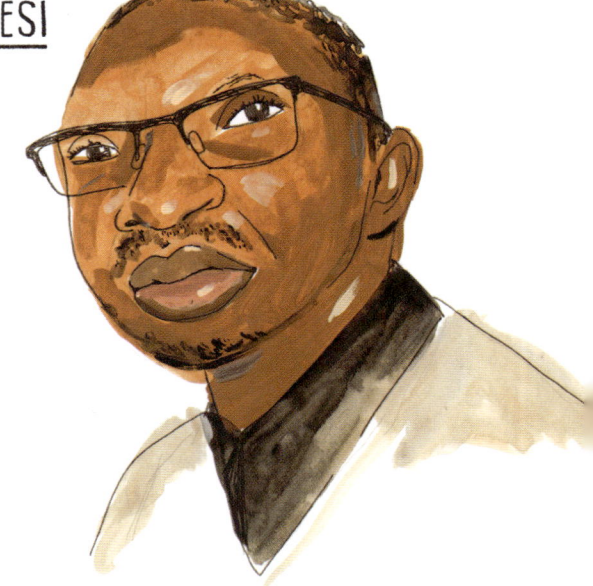

Dr. Ekesi is a renowned entomologist and professor at the International Centre of Insect Physiology and Ecology (ICIPE) in Kenya.

He specializes in integrated pest management and has made significant contributions to sustainable agriculture through his research on insect pests and how to control them. He has been recognized with The World Academy of Sciences (TWAS) Award in Agricultural Sciences and the Distinguished Scientist Award from the Entomological Society of America. He was kind enough to answer a few questions.

JR/MH: YOU'VE LED AN EXTREMELY SUCCESSFUL FRUIT FLY CONTROL INITIATIVE IN AFRICA. TELL US MORE.

SE: Damage caused by fruit flies can result in 40–80 percent losses to fruit and vegetable harvests. With my team, I have developed and implemented alternative pest management techniques that mitigate environmental risk and human health effects of insecticide use. The program has benefited over 100 million households across Africa, transforming agrifood systems, improving economic well-being, and revolutionizing horticultural crop production for both domestic and export markets.

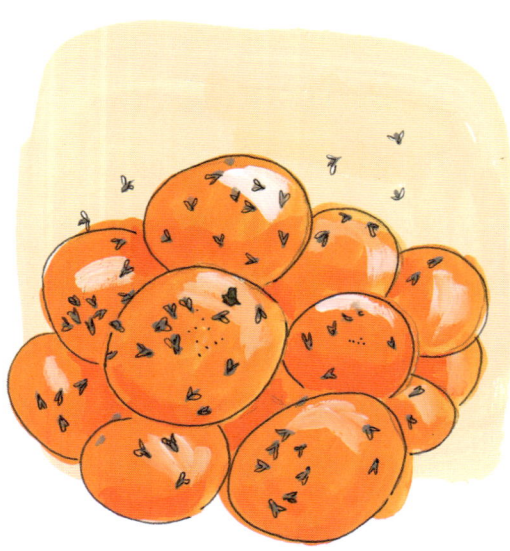

DESCRIBE THE "INSECTS FOR FOOD AND FEED" PROGRAM THAT YOU INITIATED.

Insects, which are a traditional food source for over 2 billion people globally and are widely consumed in Africa, offer a sustainable alternative but are typically harvested from the wild with environmental drawbacks. My team has developed innovative techniques for the mass production of edible insects, influencing policy and creating standards that support their use in food for both humans and animals. This has enabled significant production of insect-based protein in Kenya, benefiting poultry and aquaculture industries while also providing organic fertilizer to enhance soil health and reduce the need for synthetic fertilizers.

WHAT ARE SOME OF THE BIGGEST CHALLENGES YOU FACE AS AN ENTOMOLOGIST?

Global challenges like population growth, climate change, and invasive species mean that entomologists often must compete with scientists in other fields for resources to tackle insect-related issues. When invasive species emerge, entomologists frequently have to react to damage after it has occurred. Furthermore, the number of entomologists worldwide is declining, highlighting the need to strengthen expertise in this field.

" Only a fraction of the over 5 million known insect species are pests. It can be difficult to convince the public of the crucial role insects play in supporting ecosystems."

WHAT MOTIVATES YOU IN YOUR RESEARCH AND LEADERSHIP ROLES? HOW DO YOU STAY INSPIRED AMIDST CHALLENGES?

While this work is challenging, it is achievable and rewarding. I am especially motivated when I see end-users applying the research tools and technologies in their daily lives and making a positive impact. Additionally, I find great satisfaction in training the next generation of entomologists, many of whom stay in Africa to advance the field with the same enthusiasm and passion.

INSECT FOSSILS

Dinosaurs and woolly mammoths may be the first fossils most people think of, but fossilized insects are particularly important to scientists who study changing climate and evolution.

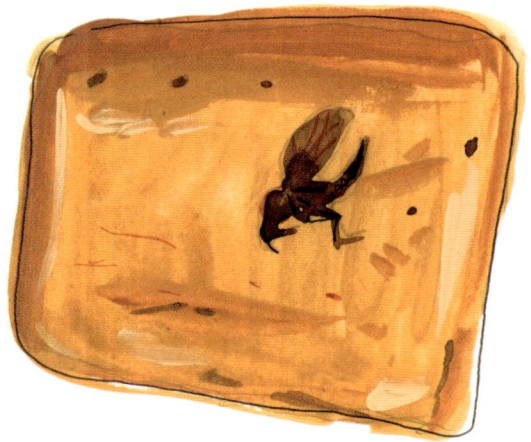

Although the oldest insect fossils date back more than 400 million years, the majority of discoveries are from the Jurassic Period, approximately 200 million years ago. In some instances, impressions are discovered within sedimentary rock, while in other cases, entire insects have been found nearly perfectly preserved in hardened tree resin, known as amber.

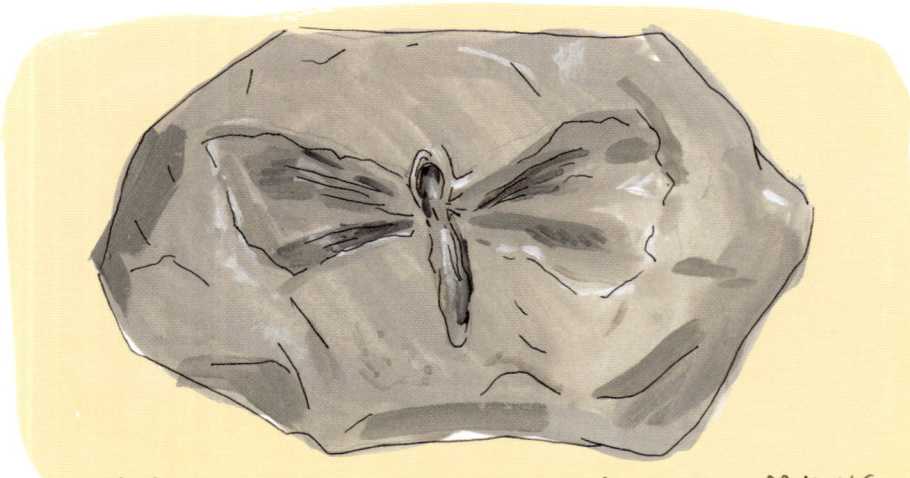

MEGANEURA MONYI AT MUSEUM OF NATURAL SCIENCES, BRUSSELS

Beetles, flies, dragonflies, and true bugs are some of the more notable insects from the Jurassic Period that have been found fossilized in various deposits around the world.

CRITICALLY ENDANGERED

Like many creatures on Earth, numerous species of insects are critically endangered cue to habitat loss, pollution, climate change, and other factors.

Environmental organizations such as the global authority the International Union for Conservation of Nature (IUCN) evaluate the risk of a species becoming extinct and categorize it. They look at population size, geographic range, rate of decline, disruption of habitat, and other factors. Species get labeled with designations like:

| EXTINCT | EXTINCT IN THE WILD | CRITICALLY ENDANGERED | ENDANGERED | VULNERABLE | NEAR THREATENED | LEAST CONCERN |

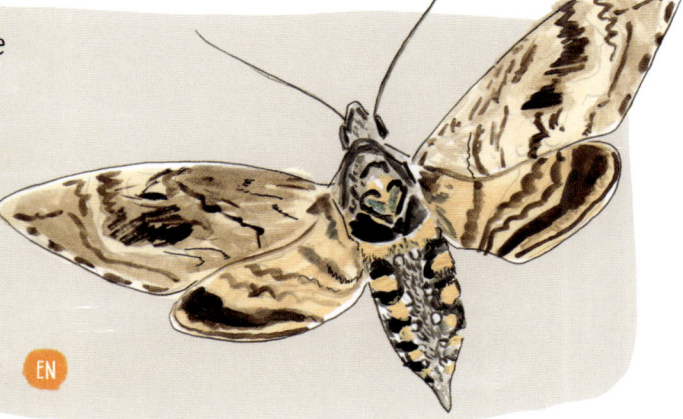

One of Hawaii's largest native insects, Blackburn's sphinx moth, which has a wingspan of up to 5 inches, was thought to be extinct until 1984 when it was rediscovered. Nonetheless, they are in decline due to habitat loss and newly introduced plants and animals. **EN**

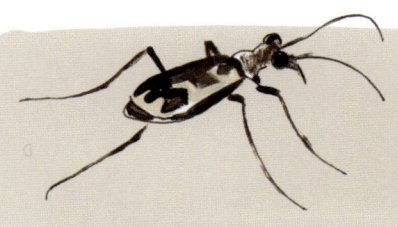

Found only on the saline wetlands of Nebraska, the Salt Creek tiger beetle is one of the rarest and most endangered insects in the United States. **CR**

Another insect in rapid decline due to habitat loss, particularly in the Midwest of the United States, is the Karner blue butterfly. **CR**

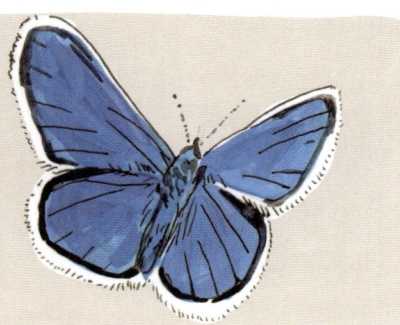

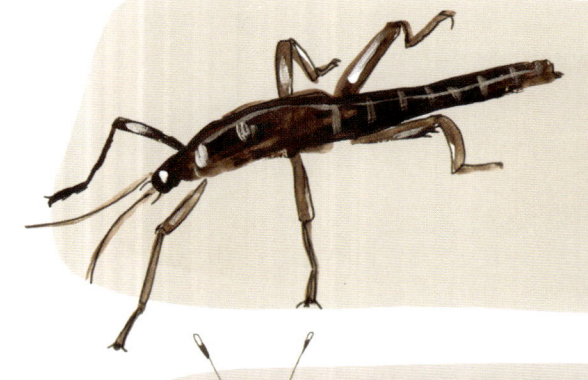

The Lord Howe Island stick insect of Australia was thought to have become extinct soon after rats were introduced to the island in 1918. A tiny population discovered in 2001 formed the basis for a captive breeding program. **CR**

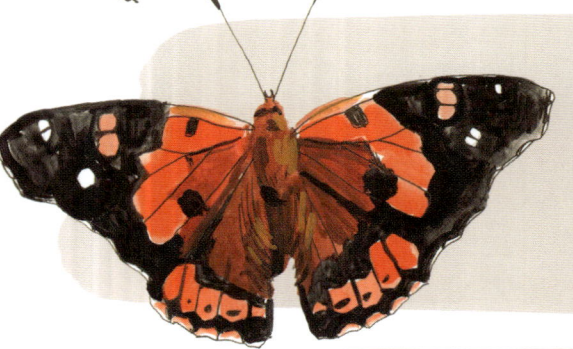

The Kamehameha butterfly, Hawaii's state insect, endemic to the state and found on all four major Hawaiian islands, is now considered a vulnurable species due to forest habitat loss and non-native species predation. **VU**

The Delhi Sands flower-loving fly has been endangered since 1993. Found in Southern California and threatened by the sprawl of newly-constructed homes, buildings, and roads, it is currently the only fly to receive protected status under the Endangered Species Act. **EN**

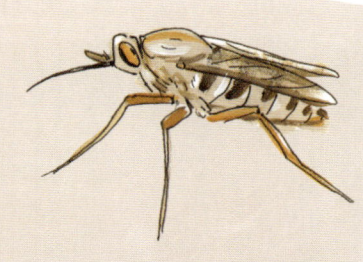

FIVE THINGS TO HELP INSECTS

We need insects. Unfortunately, there is substantial evidence indicating that insect populations are declining globally. This raises concerns about the potential consequences for ecosystems, food security, and human well-being. In addition to creating pollinator- and wildlife-friendly areas where you live, here are a few things you can do to help save insects.

1. SUPPORT ORGANIC FARMING

Choose organic produce and support organic farming practices, which promote healthier ecosystems for insects.

2. REDUCE OUTDOOR LIGHTING

Light pollution can disrupt nocturnal insects, such as moths and fireflies, which rely on darkness for navigation and mating.

3. ENGAGE IN INSECT-FRIENDLY GARDENING PRACTICES

Plant native species, let your grass grow, embrace weeds, don't rake leaves, and don't spray pesticides anywhere around your property.

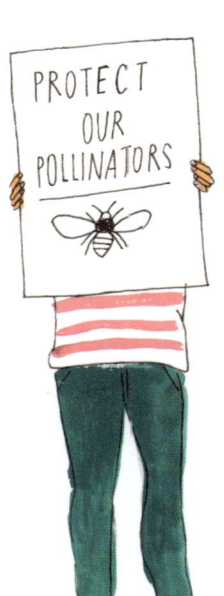

4. SUPPORT INSECT-FRIENDLY POLICIES

Get involved with politics. Advocate for policies that prioritize insect conservation, sustainable and management, recycling, and efforts to combat climate change. Vote!

5. EDUCATE OTHERS

Spread awareness about the importance of insects. Encourage friends, family, and community members to join in efforts to protect and conserve insects and their habitats.

GO ON A BUG HUNT

There is no better way to learn about insects than to step outside and search for various species within their natural habitats. You can explore areas like gardens, parks, or forests, carefully observing insect behaviors and their roles in the ecosystem while enjoying the thrill of discovery.

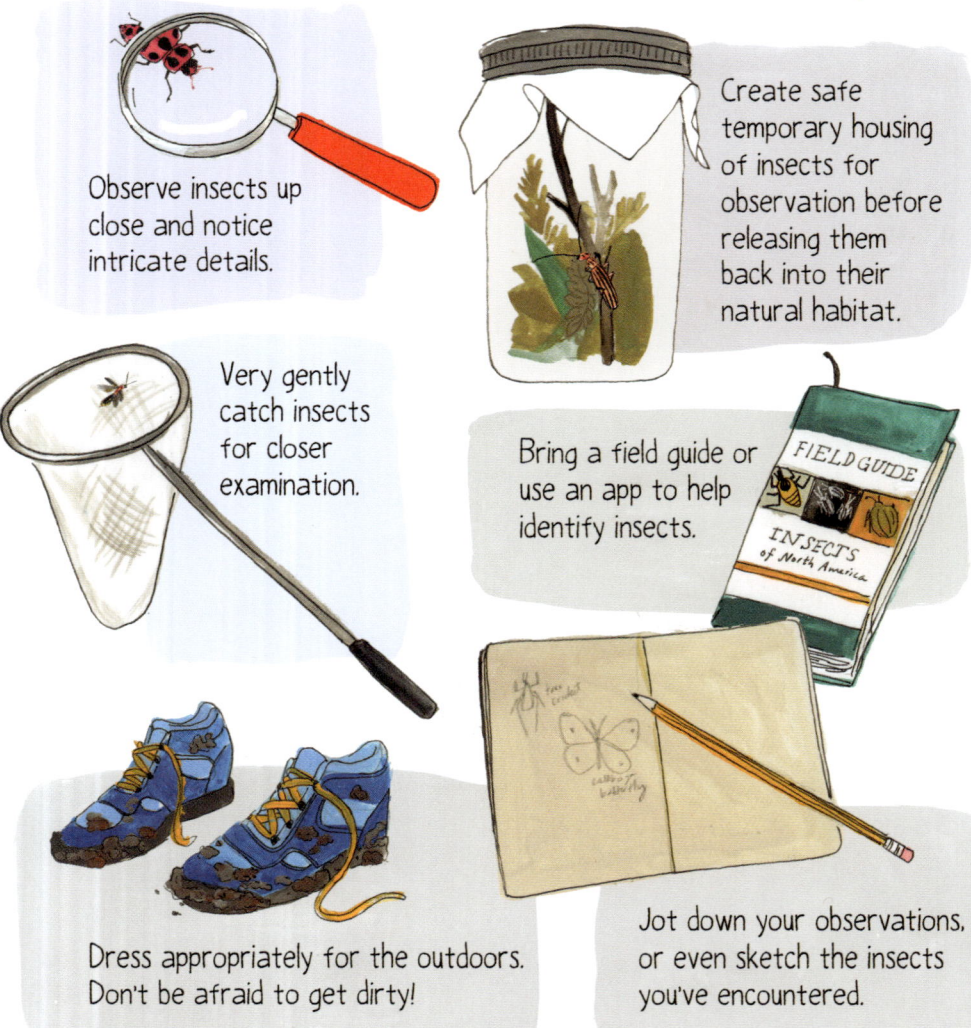

Observe insects up close and notice intricate details.

Create safe temporary housing of insects for observation before releasing them back into their natural habitat.

Very gently catch insects for closer examination.

Bring a field guide or use an app to help identify insects.

FIELD GUIDE

INSECTS of North America

Dress appropriately for the outdoors. Don't be afraid to get dirty!

Jot down your observations, or even sketch the insects you've encountered.

BUILD A BUG HOTEL

Insects appreciate a safe space to hang out in and raise their young. You can help by building a bug hotel in your backyard or garden area. Lots of different kinds of insects will come to dwell and you'll be able to easily observe them.

Use organic materials like wood and bricks to create a structure with divided layers and sections.

Fill the spaces with pine cones, grasses, bark, brush, twigs, and hollow stems.

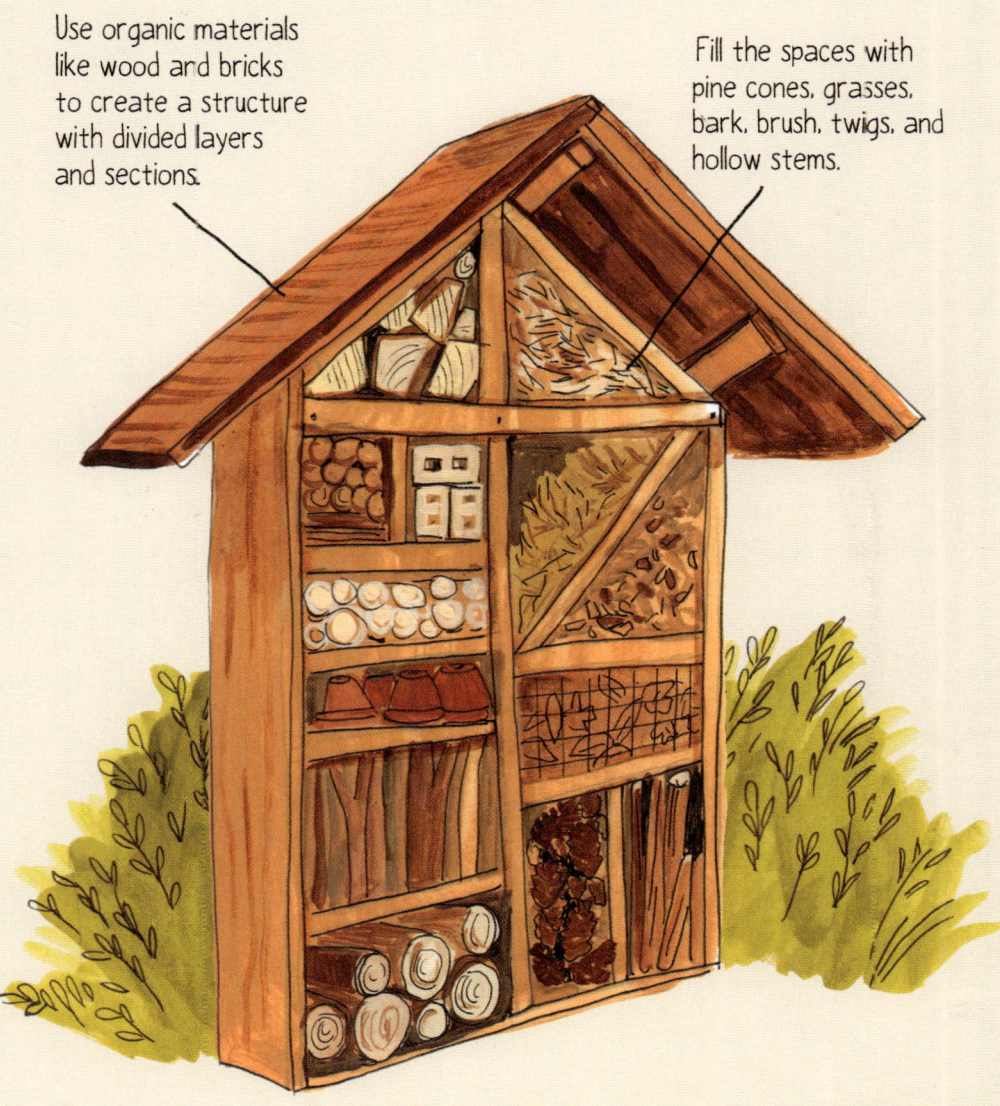

PLANT A BUTTERFLY GARDEN

PLANT FLOWERS

Some popular butterfly-attracting flowers include milkweed, lantana, verbena, and marigold, as well as the ones shown below. Research which butterflies are native to your area and include their preferred host plants where they lay their eggs. For example, monarchs need milkweed for food and egg laying.

CREATE PUDDLING AREAS

Butterflies congregate in moist areas to drink water. You can create a butterfly puddling area by filling a shallow dish with sand or soil and keeping it moist. Adding a few rocks or pebbles for butterflies to perch on can make it even more attractive.

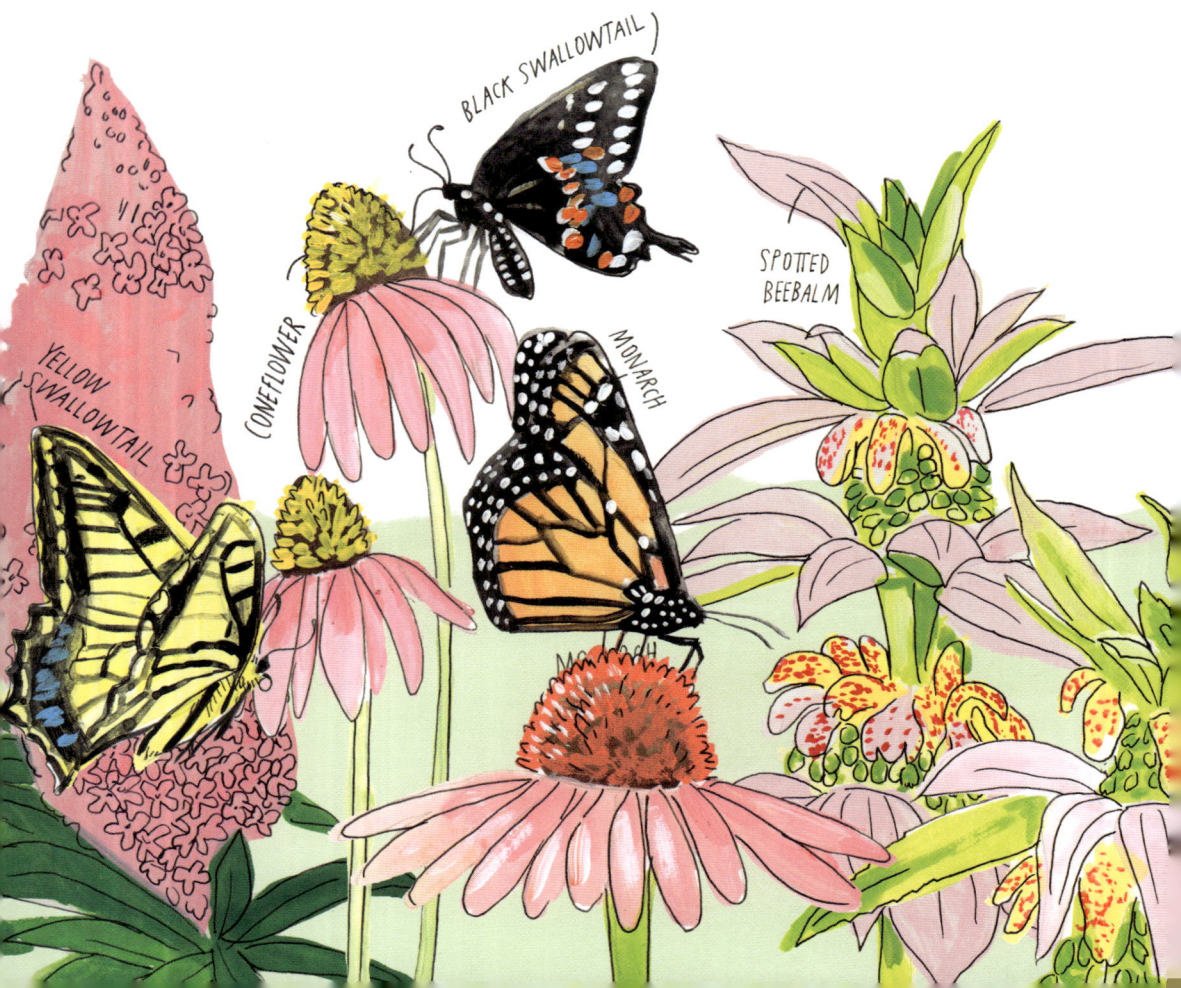

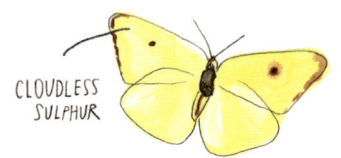
CLOUDLESS
SULPHUR

SUNNY SPOTS

Butterflies are ectothermic,
meaning they rely on external
sources of heat to regulate
their body temperature. In
sunny spots in your garden,
provide resting surfaces
where butterflies can bask.

AVOID PESTICIDES

Instead of using chemical
pesticides, opt for organic pest
control methods or encourage
natural predators to keep pest
populations in check.

GO WILD

Minimize disturbing the butterfly's
habitats by avoiding excessive pruning
or mowing. Allow some areas of your
garden to grow wild to provide
habitat for caterpillars and pupae.

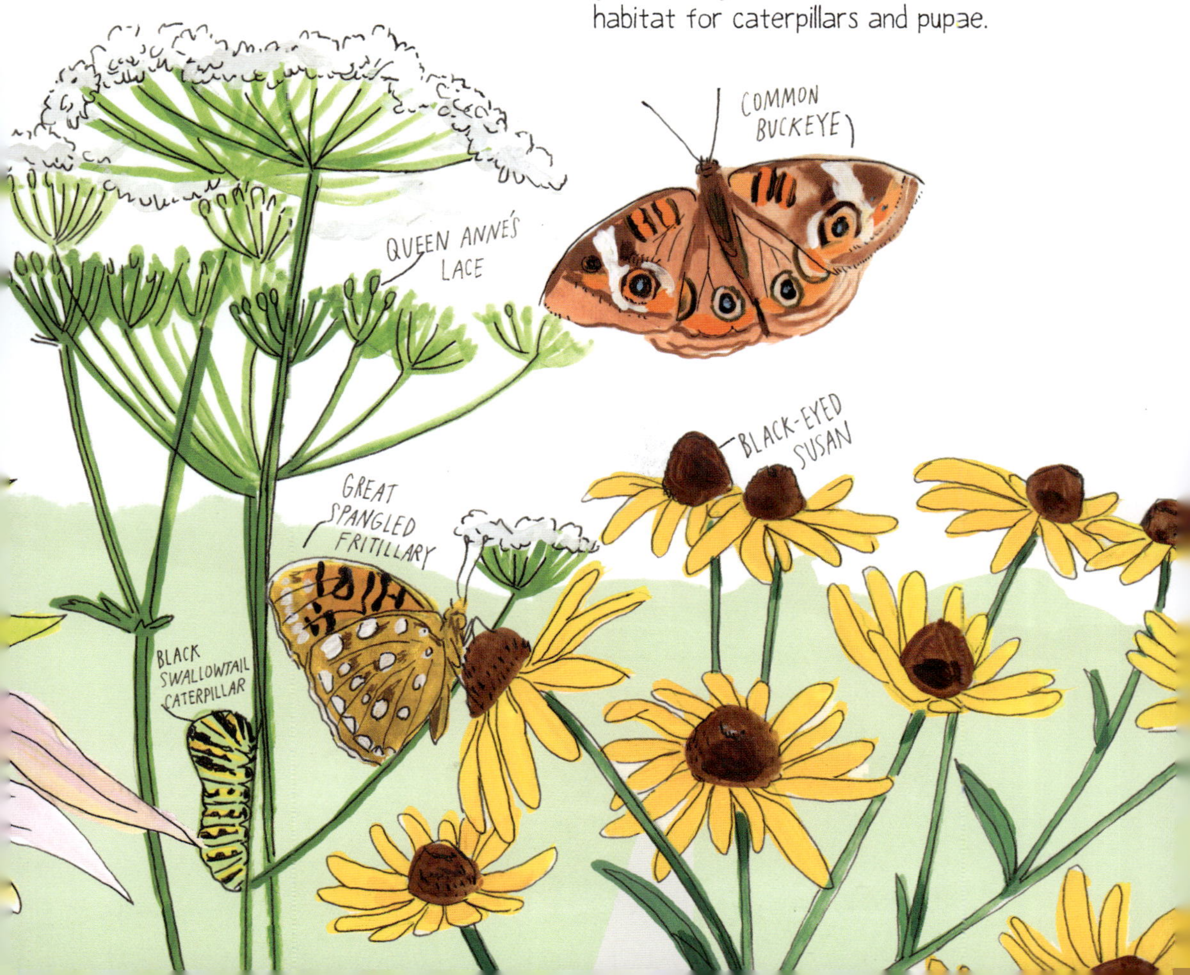

QUEEN ANNE'S
LACE

COMMON
BUCKEYE

BLACK-EYED
SUSAN

GREAT
SPANGLED
FRITILLARY

BLACK
SWALLOWTAIL
CATERPILLAR

RECOMMENDED READING & MORE

JULIA RECOMMENDS...

I loved the book *Insects of the World* from Australian entomologist and photographer Paul Zborowski. The photos are amazing, the depth of information extensive.

There are some incredible insect photographers sharing their work on Instagram. I especially love following Alejo Lopez @junglediamonds to see up-close images of every kind of insect.

For my birthday, my husband, Ollie, got me a membership to the Xerces Society, "an international nonprofit organization that protects the natural world through the conservation of invertebrates and their habitats."

Lastly, if you're ever in New York City, the American Museum of Natural History has a facinating insectarium where you can watch leafcutter ants at work.

MICHAEL RECOMMENDS...

While Julia's illustrations are truly magnificent, I was stunned by the beauty of Joel Sartore's work in *National Geographic's Photo Ark Insects*. On that note, Nat Geo also has many wonderful online videos. I particularly enjoyed the *Incredible Critters!* series.

The library!! Okay, ignore what I said about the internet and visit your local library. Maybe they even still have a card catalog or microfiche . . . though that's doubtful. Nonetheless, you will easily find books on insects.
Hey, maybe you even found THIS book in your library?

As we mentioned in the final chapter, go on a bug hunt. See how many different insects you can identify. Take a guidebook with you such as the *National Audubon Society Field Guide to Insects and Spiders*. If you want to be fancy, you can use an app like *Picture Insect*, which allows you to use your smartphone's camera to help with the identification, though that's sort of cheating.

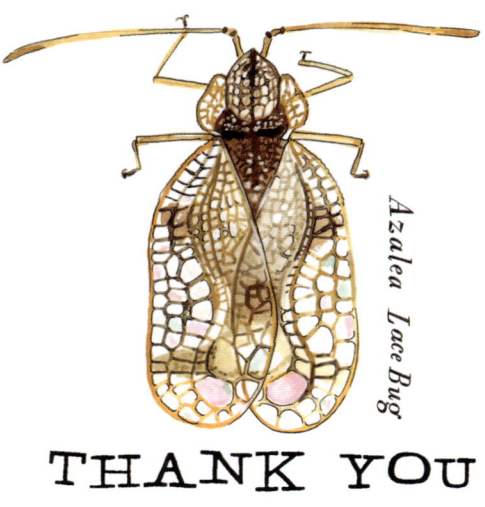

Azalea Lace Bug

THANK YOU

Michael Hearst, who lives only a few blocks away and would meet up with me over tea to talk about all the fascinating facts he found for our book. Check out his other books, especially *Unusual Creatures*, which has an album of songs to accompany it!

Lisa Hiley, our editor, whom we love and who has worked with me for so many years now on the entire series.

Deborah Balmuth, Margaret Lennon, Alethea Morrison, HK Goldstein, Alee Moncy, Maddy Jackson, Jennifer Travis, Melinda Slaving, and the entire staff of Storey Publishing, who are a huge pleasure to work with.

The Beetlelady (www.beetlelady.com), Stephanie Dole, who fact-checked these pages.

Contributors to my sister, Jessica Rothman's African Wildlife Conservation Fund for Ugandan Student Graduate Degrees and Research: Lee Ann Adams, Daniel Cochran, Shaina Feinberg, Brett Irizarry, Caitlin Keegan, Jackson Krule, Joanna Laine, Matthew Leines, Nina Lauro, Danielle McGurran, Katie Strauss, Mariah Walker, and Emily Wheeler.

Ollie, for his continued help, support, and love.

All the kids (and adults!) who have written to me! What Anatomy book would you like to see next? Find me on Instagram @juliarothman to follow everything I'm working on.

A NOTE ABOUT OUR PROCESS

I created all the artwork for this book using Holbein Acryla Gouache, outlining each piece with ink. Unlike the previous books in this series, where I separated the line work and painting and combined it digitally, this time I drew and painted together. This approach allowed for more intricate details than in my earlier works. After scanning the paintings, I designed the layouts in Photoshop. I also hand-lettered some of the titles and used fonts I created (including this one!) for the rest of the text. I thoroughly enjoyed this process and plan to continue with it—especially now that I have a stack of original paintings nearly six inches tall in my studio, which I hope will eventually find their way into people's homes.

For the insect references, I relied primarily on WikiCommons, a royalty-free image-sharing site. I am deeply grateful to all the photographers who generously share their images with the public.

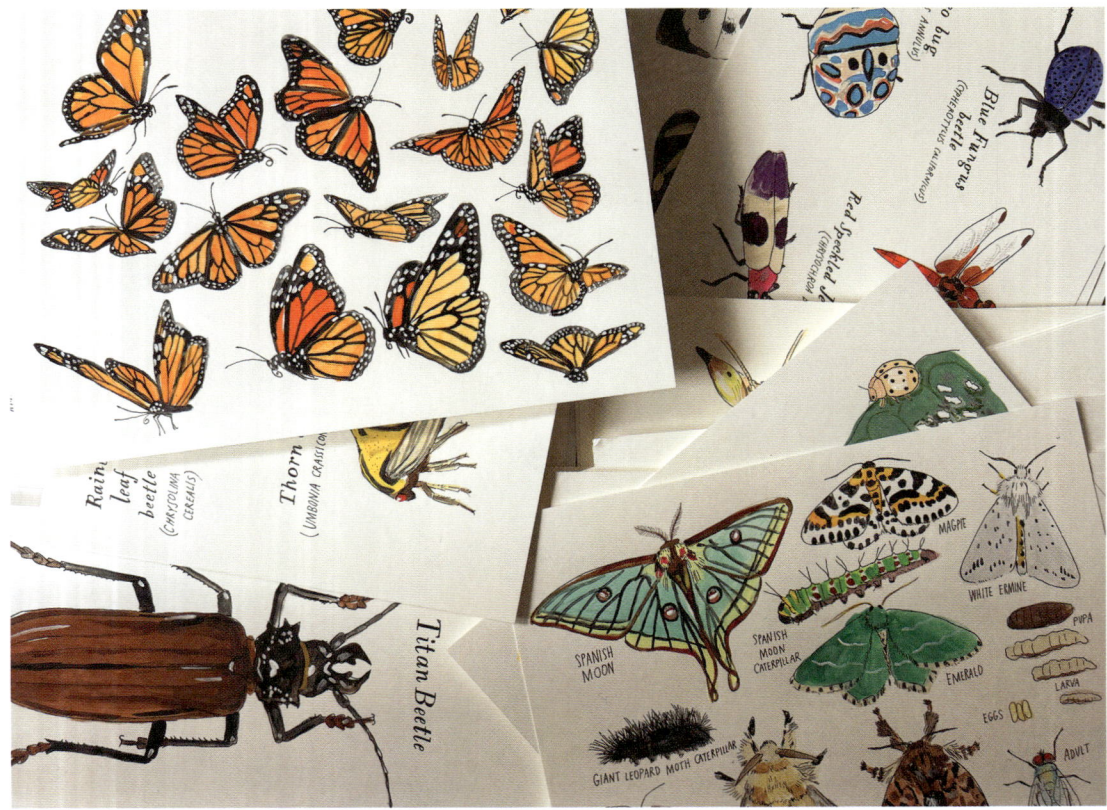

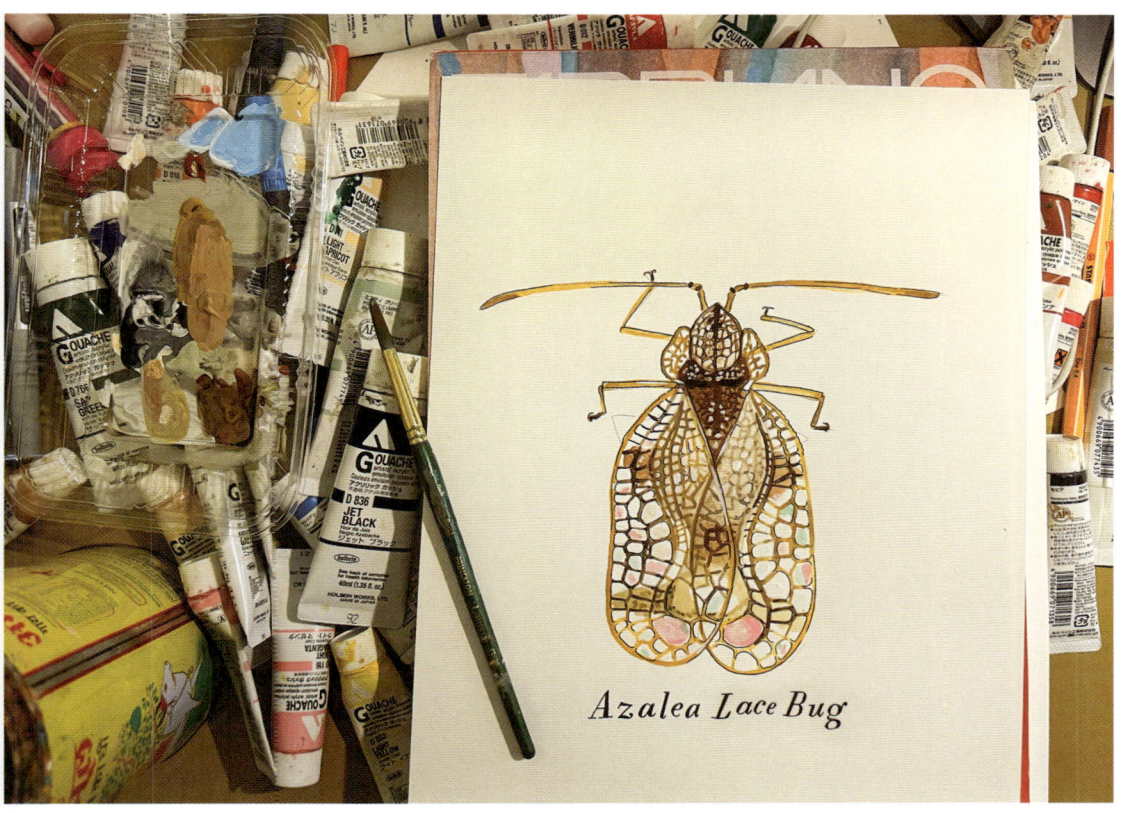

Azalea Lace Bug

I collaborated with my editor, Lisa Hiley, to create an outline for the book that detailed everything we wanted to cover. While I painted, Michael and Lisa worked on the text. Along the way, we often stumbled on new topics and ended up including many things we hadn't originally planned.

The entire process of creating a book takes me about a year, followed by additional time for editing, fact-checking, printing, and shipping. I started drawing the insects for this book in August 2023 and submitted the first draft by the end of August 2024.

With a limit of 208 pages, this book can only scratch the surface of the vast world of insects. As with all the books in this series, I feel incredibly fortunate to continue learning through the process of creating.

COLLECT THE WHOLE

BOOKS

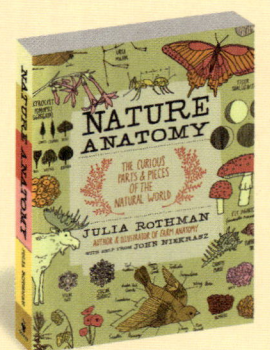

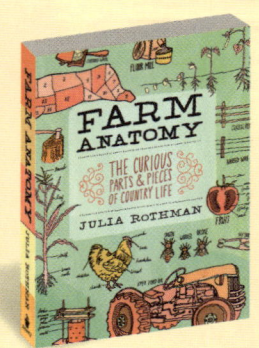

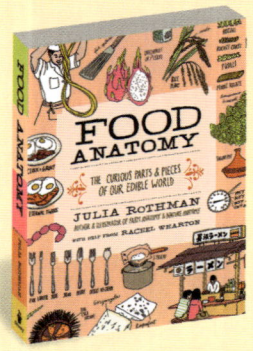

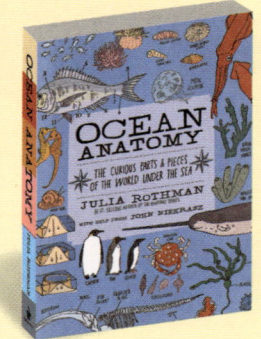

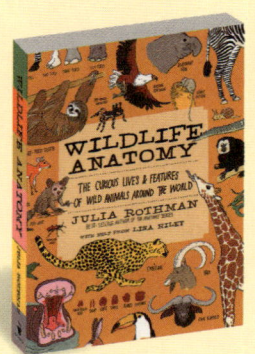

NOTEBOOK

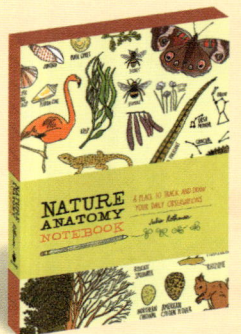

BOXED SET

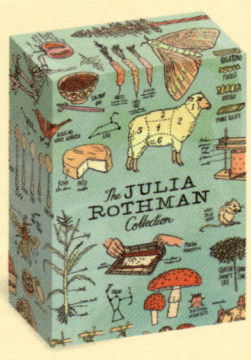

ANATOMY SERIES!

ACTIVITY BOOKS

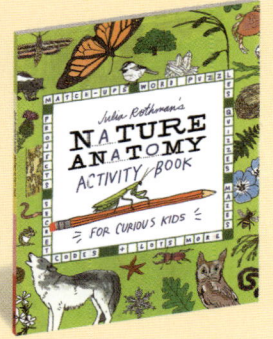 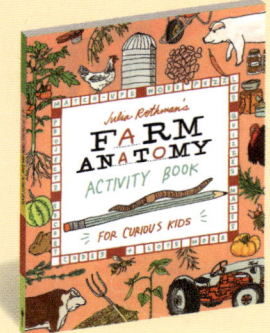 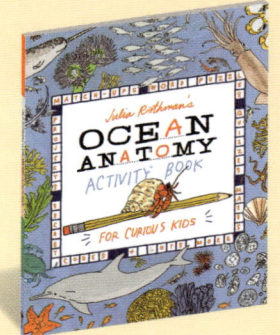

STICKER BOOKS

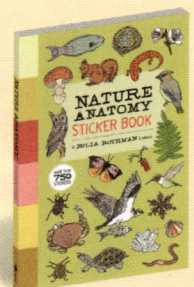 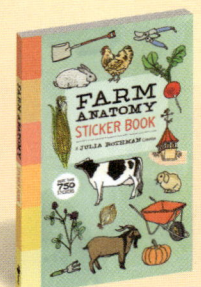 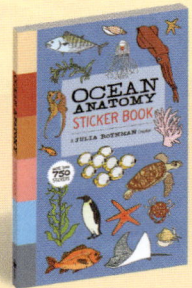 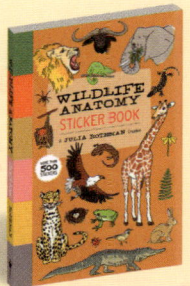

PUZZLES

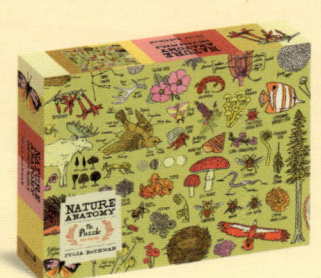 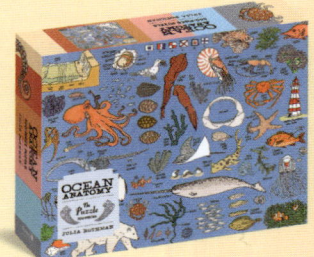 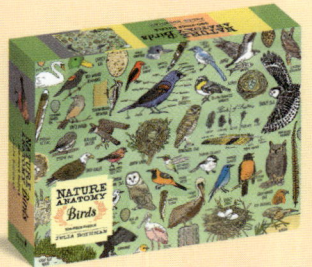

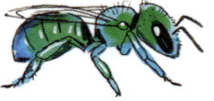